Investigate the Possibilities
W9-AAJ-371
Elementary Chemistry
MATTER
Its Properties & Its Changes
Tom DeRosa
Carolyn Reeves

Tom DeRosa
Carolyn Reeves

First Printing: April 2009

Master Books®
P.O. Box 726
Green Forest, AR 72638

Printed in the United States of America

Cover Design by Diana Bogardus
Interior Design by Terry White

ISBN 10: 0-89051-560-3
ISBN 13: 978-0-89051-560-0
Library of Congress number: 2009923588

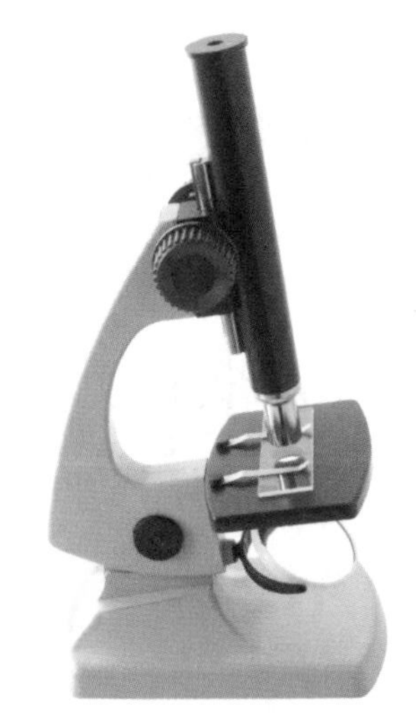

All Scripture references are New International Version unless otherwise noted.

Please visit our website for other great titles: www.masterbooks.net

TABLE OF CONTENTS

About the Authors

Tom DeRosa, as an experienced science educator and a committed creationist, has incorporated both his passions in the founding and the directing of the Creation Studies Institute, a growing national creation organization whose chief focus is education. His wealth of experience in the public school classroom, Christian school, and the homeschool markets for over 35 years has given special insights into what really works in engaging young minds. He holds a masters degree in education, with the emphasis of science curriculum. He is an author and sought-out, enthusiastic creation speaker who has a genuine love for the education of our next generation.

Carolyn Reeves is especially skilled at creating ways to help students develop a greater understanding of not just scientific concepts, but also how these are applied within the world around us. Carolyn retired after a 30-year career as a science teacher, finished a doctoral degree in science education, and began a new venture as a writer and an educational consultant. She and her husband make their home in Oxford, MS, where they are active members of North Oxford Baptist Church. The Reeves have three children, three in-law children, and ten grandchildren.

Photo Credits

L = left, TL = top left, BL = bottom left, R = right, TR = top right, BR = bottom right, TC = top center, BC =bottom center

All photos are from Shutterstock unless otherwise noted.

Clipart.com: pg. 4-R, pg. 17-TR, pg. 48-L, BL, pg. 54-LC.

Created by Design: pg. 71-BR, pg. 73-BR.

Faith Dipasquale: pg. 30-R, 61TR and BL.

Istock.com: pg. 43-BR, pg. 60-TL, pg. 83-BR, pg. 84-BR, pg. 85.

Master Books Inc.: pg. 29-TC, pg. 87

NASA: pg. 41.

Wikimedia: pg. 30-L, pg. 50-L, pg. 54-L, pg. 80-L, pg. 82-L, pg. 85-R.

INTRODUCTION

In this book, we will be studying matter, its properties, its changes, and its interactions. The investigations will include things you can observe or measure. However, much that is going on will involve tiny particles you cannot observe. Have you ever wondered how scientists know so much about things they cannot see?

You will come to appreciate the creativity of the early scientists who were able to understand so many things about matter. Many of these scientists had full-time jobs and did their research in their spare time.

Sometimes early scientists had to undo wrong ideas. One ancient theory stated that there were only four basic elements — water, air, earth, and fire. It was wrong, of course, but it was an idea that stayed around for almost 2,000 years.

Some of the first chemists were known as alchemists. They thought they could make gold out of lead and other substances. They were also wrong, but scientists still learned some important ideas about science through them.

The science of chemistry came about in spite of false starts and mistakes. There were times when scientists made important discoveries, but didn't realize they had done so. There were frequent times when scientists vigorously disagreed with each other about their explanations and interpretations.

New knowledge emerged at a slow pace, with clues coming from many scientists. They started with things they could observe. They learned more about the properties of substances and how they changed under different conditions. They gradually came to realize that matter is made up of tiny particles too small to be seen.

New technologies have led to even more amazing discoveries about the chemical substances around us. There is still much to learn and understand about the invisible world of atoms and molecules in the future. Perhaps future chemists will help solve some of the world's big problems by producing alternative fuels, food sources, building materials, or medicines.

You may be surprised to learn that a large number of the early chemists were dedicated Christians who considered their research a way to bring glory to God the Creator. One of the Christian chemists from the 1600s (Robert Boyle) wrote, "When . . . I study the book of nature I find myself oftentimes reduced to exclaim with the Psalmist, How manifold are Thy works, O Lord! In wisdom hast Thou made them all!" We hope that you too will recognize the great wisdom of our Creator as you investigate the world of matter.

Scientists

From 1600s:	Robert Boyle	(1627–1691)
From late 1700s:	Henry Cavendish	(1731–1810)
	Antoine Lavoisier	(1743–1794)
	Joseph Priestley	(1733–1804)
From early 1800s:	John Dalton	(1766–1844)
From late 1800s:	Dmitri Mendeleev	(1834–1907)
From early 1900s:	William Thomson (Lord Kelvin)	(1871–1937)

HOW TO USE THIS BOOK

Each investigation gives students a chance to learn more about some part of God's creation. To get the most out of this book, students should do each section in order. Many science educators believe science is best learned when students begin with an investigation that raises questions about why or how things happen, rather than beginning with the explanation. The learning progression recommended for this book is: engage, investigate, explain, apply, expand, and assess. In each lesson, students will be introduced to something that is interesting, they will do an investigation, they will find a scientific explanation for what happened, they will be able to apply this knowledge to other situations and ideas, they will have opportunities to expand what they learned, and there will be multiple assessments.

Think about This – (Engage) Students should make a note of what they know or have experienced about the topic. If this is a new topic, they could write some questions about what they would like to learn.

The Investigative Problem(s) – Students should be sure to read this so they will know what to be looking for during the investigation.

Gather These Things – Having everything ready before starting the investigation will help students be more organized and ready to begin.

Procedures and Observations – (Investigate) Students should first follow the instructions given and make observations of what happens. There will usually be opportunities for students to be more creative later.

The Science Stuff – (Explain) This section will help students understand the science behind what they observed in the investigation. The explanations will make more sense if they do the investigation first.

Making Connections – (Apply) Knowledge becomes more permanent and meaningful when it is related to other situations and ideas.

Dig Deeper – (Expand) This is an opportunity for students to expand what they have learned. Since different students will have different interests, having choices in topics and learning styles is very motivating. All students should aim to complete one "Dig Deeper" project each week, but the teacher may want older students to do more. Generally, students will do at least one project from each lesson, but this is not essential. It is all right for students to do more than one project from one lesson and none from another.

What Did You Learn? – (Assessment) The questions, the investigations, and the projects are all different types of assessments. For "What Did You Learn?" questions, students should first look for answers on their own, but they should be sure to correct answers that might not be accurate.

The Stumpers Corner – Students should ask classmates (or others) two original short-answer questions about what they have learned in this lesson, noting if the questions stumped anyone or not. Write the questions and answers in the Student Answer Book. Another option is to write two questions they would like to learn more about.

Additional opportunities for creative projects and contests are found throughout the book. For grading purposes, they can be counted as extra credit or like a "Dig Deeper" project.

Nurture Wisdom and Expression

Each book contains information about early scientists and engineers. Students need to see that they were regular people who had personal dreams and who struggled with problems that came into their lives. Students may be surprised to realize how many of the early scientists believed that understanding the natural world gave glory to God and showed His wisdom and power.

In addition to the science part, students will find creation apologetics and Bible mini-lessons. The apologetics will clear up many of the misconceptions students have about what science is and how it works. Both the apologetics and Bible lessons should lead to worthwhile discussions that will help students as they form their personal worldviews.

Students with artistic and other creative interests will have opportunities to express themselves. For example, some of the apologetics are written in narrative form and are suitable for drama presentations. As scientists are introduced and researched, students can also present what they have learned as time-dated interviews or news accounts. Remember, if the scientists are included in a drama presentation, they should be represented as professionals, not as stereotyped, weird-looking people.

These experiments require adult supervision. They have been specifically designed for educational purposes, with materials that are readily available. At their conclusion, please appropriately dispose of any by-products or food items included in the experiments.

The Physical Side of Chemicals

Think about This

A detective collected samples of food from the table where a victim was eating when he collapsed. The detective sent them to a crime lab. A few days later, the lab called to say they had positively identified a poison in the victim's food that was not in anyone else's food. Have you ever wondered how someone in the crime lab could figure out what chemicals are present in food or in someone's blood or in something else?

The Investigative Problems

How can the physical properties of a chemical substance be used to help identify the substance?

Procedure & Observations

Your teacher will show you ten items. Your job is to identify one of the items on the basis of its physical properties. You should eliminate any item that doesn't match the descriptions. These are the physical properties of the item: It is round. It is flat. You would not want to eat it. It would be hard to break. It is shiny. What is the item that has all of these properties?

Your teacher will give you some more substances to investigate, but each of these will be a pure chemical substance. They will be either an element or a compound.

Bring a magnet near each substance and observe if the magnet has an effect on it. Place each substance in a container of water and observe if it floats or sinks. Note if it is soluble (will dissolve) or insoluble (will not dissolve) in the water. Note also the color and whether it is shiny or dull. Put this information in a data table.

Substance	Effect of a magnet	Float or sink in water	Soluble or insoluble in water	Color	Shiny or dull
Iron nail					
Paraffin					
Sugar cube					
Oil					
Copper penny					

Use your chart to identify each substance.

1. Which substance is attracted to a magnet?
2. Which substance is a shiny orange-brown color and sinks in water?
3. Which substance is soluble (dissolves) in water?
4. Which substance is a solid and floats on water?
5. Which substance is not a solid and floats on water?

The Science Stuff

Physical properties are often characteristics you can see, hear, taste, smell, or feel, but may include any physical characteristics of a substance. You used some simple physical characteristics to identify one of the ten items you were first shown.

Some of the items you were shown were pure substances (like the glass) and some were a mixture of many substances (like the apple). A pure chemical substance could be either an element or a compound. (We'll learn more about elements and compounds later.) A fragment of a pure substance would have the same properties as the whole substance. All of the basic particles in a pure substance are the same. For example, a piece of pure iron only contains particles of iron and a container of pure water only contains particles of water.

Properties such as size and shape were helpful in identifying the first items, but they are seldom considered in identifying pure chemical substances. The properties of the five pure substances listed in the chart will be present regardless of the size, shape, or amount of the substance. Scientists look for characteristics that will remain the same no matter where the chemical is found. Almost any substance can be made into a round shape, so this would not be helpful in knowing what chemical is present.

We examined physical properties of several pure substances, including the effects of a magnet, whether the substance would float or sink in water, whether the substance was soluble or insoluble in water, its color, and its shininess. There are many other properties we could have considered, such as odor, taste, density, hardness, brittleness, elasticity, melting and boiling temperatures, solubility in other liquids, conductivity of heat and electricity, and viscosity.

Making Connections

There is a huge need for methods, instruments, and trained people to identify chemical substances that are present in things. Identifying unknown chemicals is part of the study of analytical chemistry. This includes what chemicals are present, their characteristics, and how much is present. There are many crime labs that hire people to help solve crimes by identifying such things as drugs, alcohol, poisons, or traces of gunpowder. Medical labs test blood and urine for the presence of many kinds of substances. Other labs help identify pollutants in the air, water, and environment. Industries must constantly monitor their products for impurities. These are only a few of the places where chemicals are analyzed.

One of the most important things any society can do is to maintain a clean source of water. During the Industrial Revolution, many factories were built next to a river so they could dump their wastes into the river. Congress eventually passed a number of laws to try and keep our water sources free of pollution. Even today, environmentalists look for better ways to prevent pesticides and other harmful chemicals from being washed into rivers and lakes after a rain.

Dig Deeper

Labs generally use both traditional methods and a variety of instruments to identify chemical substances. An instrument known as a spectroscope is often used to help analyze the chemicals in something. Do some reading about spectroscopes to find out how they work and what uses they have.

Crime labs often hire forensic scientists. What do forensic scientists do? Is there more than one kind of forensic scientist? If so, what are the different areas in which they work?

What are some of the U.S. laws that try to prevent water pollution? Do all countries have similar laws? Try to find the name of one charity whose mission is to provide clean water to people who don't have clean water to drink.

What Did You Learn?

1. What are physical properties of chemical substances?
2. When scientists want to know what chemical substances are in an item, they seldom consider the size, shape, and amount of the item. Why is that?
3. Give ten examples of physical properties used by scientists to describe a chemical substance.
4. What is a pure chemical substance?
5. What are some of the things students learn about in analytical chemistry?
6. What are some of the main things that are done in medical labs?
7. How might an environmental agency use a lab that analyzes chemical substances?
8. Are the physical properties of a piece of pure iron the same anywhere pure iron is found?

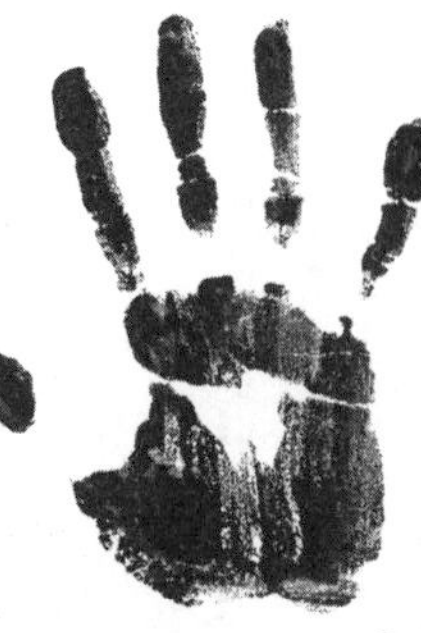

INVESTIGATION #2

Strange Substances and Their Properties

Think about This

Marita challenged her friends to guess what she had inside a plastic container. She called it MX for, Marita's unknown chemical. It moved around the bag like a liquid, but when someone squeezed it, part of it became hard like a solid. Marita asked if she could pour it into a bowl and show everyone some more properties. Her teacher agreed and said she was going to show them another really strange substance that is found in baby diapers.

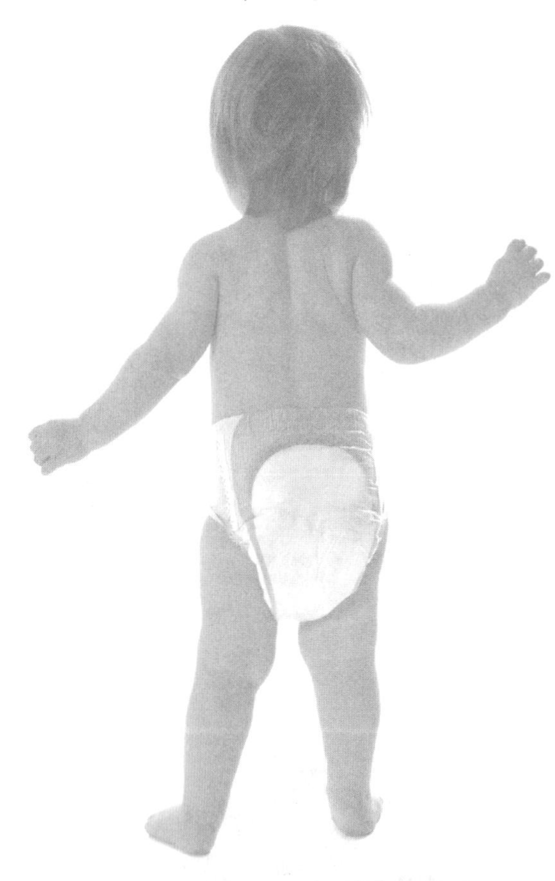

The Investigative Problems

What are the physical properties of MX and the chemical in baby diapers?

Procedure & Observations

Part I.

1. Your teacher will give you some MX in a zip bag. Look at the MX mixture through a clear zip plastic bag. Hold the bag by the different corners. Does it have properties of a liquid?
2. Hit the bag of MX (not too hard). Does it feel like a solid?
3. Pour the contents of the bag into a plastic bowl. Pick up some MX in a spoon and let it fall back into the bowl. Does the substance act like a liquid or a solid as it falls? Describe how it falls.
4. Now slowly push your finger into the MX until your finger is touching the bottom of the pan. Pull your finger out slowly. What happened?
5. Slowly push your finger into the MX again. When it is touching the bottom of the pan, try to pull your hand out quickly. What happened?
6. Now try to quickly jab the surface of the MX with your fingers. What happened?
7. Try pushing the back of a spoon through a container of MX. Move the spoon as fast as you can. Describe what happens. Now move the spoon through the MX very slowly. Is there a difference in how hard it is to push the spoon?

Part II.

1. You will need two people to do this activity. Hold a baby diaper over a pan or sink and pour 50 mL of warm water into the inside center of the diaper. Predict how much warm water you think the diaper can hold before it begins leaking. Add another 50 mL of warm water and tilt the diaper back and forth so the water can be exposed to dry areas. Continue to add 50 mL of warm water until the diaper can no longer hold any more water and it steadily leaks. Record the total amount of water you added before it began to leak. Set the diaper aside to examine later.
2. Take another diaper and separate the outer and inner lining from the middle layer. Throw away the outer stuffing and all the linings. Tear the middle layer of the diaper into small pieces. Measure the volume of these pieces of the diaper in a dry graduated cylinder or measuring cup, and record this amount. Put the pieces of the middle layer of the diaper into a gallon plastic zip bag. Add 50 mL of warm water to the bag. What do you see?
3. Continue to add 50 mL of warm water to the bag until the bag is full or the water separates from the diaper material. Keep up with the total amount of water you added.
4. Place the first diaper in a plastic bowl and pull it apart. Compare the inner contents of this diaper to the material in the gallon zip bag.
5. Estimate how much water was added for every 100 mL of dry diaper material. This doesn't need to be exact — just an estimate.
6. List some of the physical properties of the substance in the baby diaper that you observed.

The Science Stuff

MX is actually a mixture of cornstarch and water, but it acquires its own set of interesting physical properties. At times it has properties of a liquid and other times it has properties more like a solid. The water can move in and out of the cornstarch. If you press your fingers into it quickly, the water and cornstarch remains firmly in place. If you press it slowly, the mixture is very fluid. Once your fingers are in the mixture, you will have no trouble removing them slowly. It will be hard to remove them quickly.

The viscosity of MX changes under different conditions. Viscosity is a property of liquids that is related to how slowly they pour from a container or how hard it is to push something through the liquid. Molasses, for example, has a high viscosity because it pours slowly. If it is difficult to push an object through a liquid, the liquid is also said to have a high viscosity. You should have noted that you could push the spoon through the mixture slowly, but it was much more difficult to push it quickly.

Many liquids become less viscous (have less viscosity) when the temperature increases. MX may change viscosity as you apply pressure.

The chemical you separated from the baby diaper is a superabsorbent man-made polymer. It has the unusual property of being able to soak up hundreds of times its weight in water. This kind of polymer is made when many similar small chemicals (called monomers) join together to form long chains of molecules. It is similar to cotton, which is a natural polymer.

Making Connections

Having the proper viscosity is important in choosing motor oils to lubricate cars and trucks. Summer-weight and winter-weight oils allow for changes in summer and winter temperatures. Winter-weight oils are less viscous in cold weather. Summer-weight oils are more viscous in hot weather.

Temperatures reach -40° to –60° F in some places in Alaska during the winter. At these temperatures, residents must use very thin and runny motor oil (low viscosity). They must also keep a heater in the oil pan when vehicles are not in use to keep the oil warm.

The addition of water-absorbing polymers is the secret for no-leak baby diapers. Water-absorbing crystals have other exciting uses. For example, they can be placed in the soil when they are full of water (hydrated), and they will slowly release water to plant roots over a long period of time.

Dig Deeper

Talk with someone who works in a car shop that changes oil in cars. Find out more about the differences in summer-weight and winter-weight oils and why they need to be different.

Talk with someone who has lived in Alaska or another area that has long periods of freezing weather. Find out more about how they were able to keep their cars running during those very cold months.

Not too many years ago, baby diapers were made of cloth and were reusable. Today, most American parents use disposable diapers for their babies. Some environmentalists fear that these disposable diapers are creating an environmental problem. Do some research to see why they are concerned about disposable diapers.

Do an experiment by planting two groups of radish seeds. Purchase some water-absorbing polymer crystals from a lawn and garden center. (Commercial products are available under several trade names.) Let the crystals absorb a large amount of water, mix them with an equal amount of moist soil, and plant ten radish seeds in this mixture. Plant ten radish seeds in soil that does not contain the hydrated crystals. Do not water either group. For two weeks, keep a daily record of how the plants are growing.

What Did You Learn?

1. Give several physical properties of MX.
2. There are several ways to describe viscosity. Find two or more ways to describe viscosity.
3. Viscosity of oils and molasses is often affected by temperature. What affects the viscosity of MX?
4. What is one unusual property of the chemical we tested in the baby diaper?
5. What are polymers?

Chemistry Fun with Bubbles

Think about This

Mrs. Smith forgot to turn the dishwasher on before she left the house. She called home and asked someone to start the dishwasher so the dishes would be ready to use that night. The children found some liquid soap, carefully poured it into the proper containers, and turned the dishwasher on. By the time Mrs. Smith arrived home, there were several inches of soap bubbles covering the entire kitchen floor and more bubbling out of the dishwasher. What do you think the children did wrong?

The Investigative Problems

What are the physical properties of soap bubbles?

How can you make bigger bubbles?

Gather These Things:

- ✓ Water
- ✓ Straws
- ✓ Dishwashing liquid
- ✓ Glycerin
- ✓ Bubble wands
- ✓ Large container with lid
- ✓ Small plastic container for bubble solution

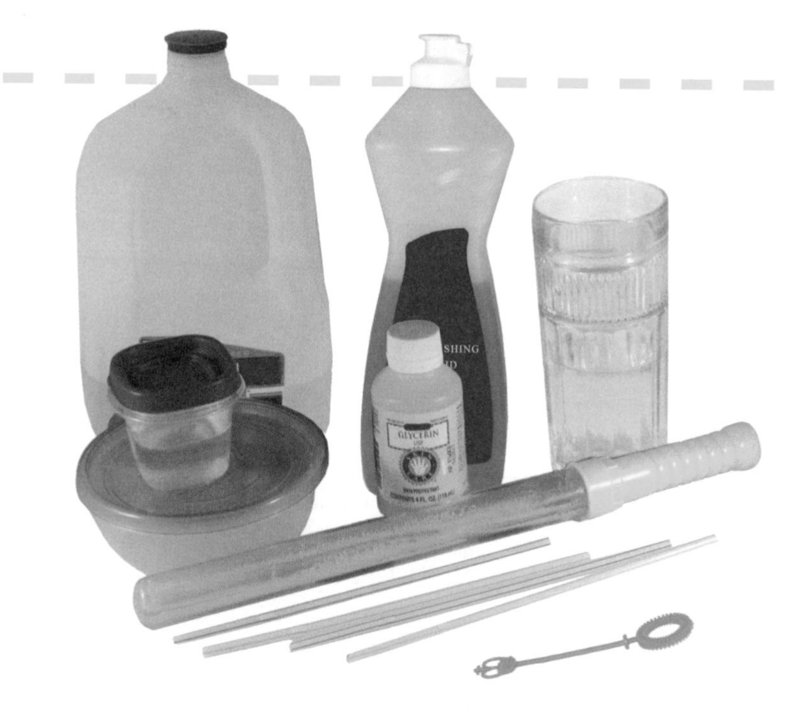

Procedure & Observations

Bubble Recipe:

Mix one cup of dishwashing liquid in one gallon of water.

Add one tablespoon of glycerin and store the solution in a container with a lid.

1. Take your bubble solution outside and use bubble wands or straws to blow bubbles.
2. See who can make the biggest bubbles.
3. See who can make their bubbles last the longest and travel the greatest distance.
4. Observe the colors of the bubbles. Notice what color they turn just before they burst.
5. Test a small amount of bubble solution without the glycerin. How does the lack of glycerin affect the bubbles?

The Science Stuff

Forcing air inside the bubbles produces pressure all around the inside of the bubble. The air pressure on the inside is balanced by the air pressure on the outside of the balloon.

All forms of matter are made up of tiny particles known as molecules. They are too small to be seen even under microscopes. An attraction between particles that are alike is called cohesion. Cohesion is an important property of matter.

There are cohesive forces between the particles (molecules) that make up the soap bubbles. These forces are especially strong at the surface of the bubbles and are known as surface tension

Not all substances have the same amount of attraction between their basic particles. The attraction between like particles (cohesion) is generally much stronger in solids than it is in liquids. It is what keeps many solid objects from falling apart. The attraction (cohesion) between gas particles is much less than that of liquids.

Another important property of a substance is its ability to return to its original shape after being stretched. This is known as elasticity. The property of elasticity is seen in soap bubbles, since they can stretch or expand quite a bit without breaking.

The water in the bubbles will eventually evaporate and cause the bubbles to break. If you watch the colors of the bubbles carefully, they will change to a gray color just before they break.

Making Connections

Soaps and detergents do not all have the same chemical makeup. They are not necessarily designed to do the same things, even though they are similar. Do you think it would be a good idea to put a bubble bath product in the washing machine to wash your clothes?

Soap does not bubble as much in hard water as it does in soft water. Hard water doesn't look or feel any different than soft water, although it may taste different. It just contains more dissolved minerals than soft water does. Some machines will include special instructions for the amount of detergent to use when clothes or dishes are washed in very hard water or very soft water.

Dishwashers need soaps or detergents that clean with low suds. Shampoo needs to be able to produce lots of bubbles from a small amount of soap or detergent.

Dig Deeper

Pour a little of the bubble recipe onto a smooth waterproof table surface. Take a straw and blow gently into the liquid soap until a bubble forms on the table. Continue blowing until it bursts. While the outline of the bubble is still visible, measure the diameter of the bubble. Devise a way to test different combinations of glycerin and dishwashing liquid to see which one produces the biggest bubbles.

You could also test different kinds of soaps or detergents to determine which brands produce the biggest bubbles.

You might want to compare the size of bubbles produced by mineral water (hard water) and distilled water (soft water). Be sure to use the same kind of soap or detergent in all your tests. Let the degree of water "hardness" be the variable.

What Did You Learn?

1. Generally, is the attraction between molecules greater in solids or in liquids?
2. Generally, is the attraction between molecules greater in liquids or in gases?
3. The attraction between molecules that are found at the surface of a liquid is called what?
4. What is the property of matter that causes like particles to attract each other?
5. What is the property of matter that allowed the bubbles to stretch without breaking (up to their limits)?
6. What is the difference between hard water and soft water?
7. Are the spherical shapes of bubbles caused more by surface tension, adhesion, or friction?

INVESTIGATION #4

Colors Are Colors

Think about This

The detective examined a note that had been handed to the bank teller during the robbery. "We have just picked up a suspect," he informed the bankers. "We're about to do a test on the ink pen in the suspect's shirt."

What kind of evidence do you think he is hoping to get?

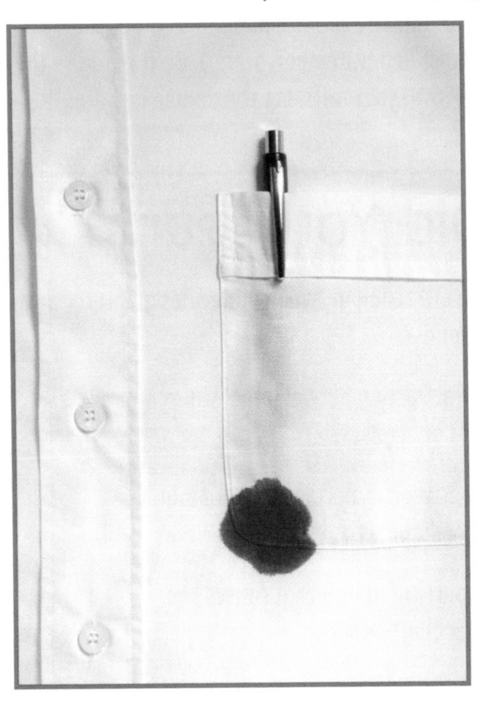

The Investigative Problems

How can you separate the chemicals in colored markers?

Procedure & Observations

Use one cup and one filter strip for each test. First, fill each cup ¼ full with water. Next, make a short line with each colored marker about ½ cm wide and 2.5 cm from the bottom edge of the strip. If more than one color is used, make a line with the first color, let it dry, and make another line with the second (and third) color directly over the first one. The mixing of colors may produce another color. Refer to the chart to see which colors to put on each filter strip.

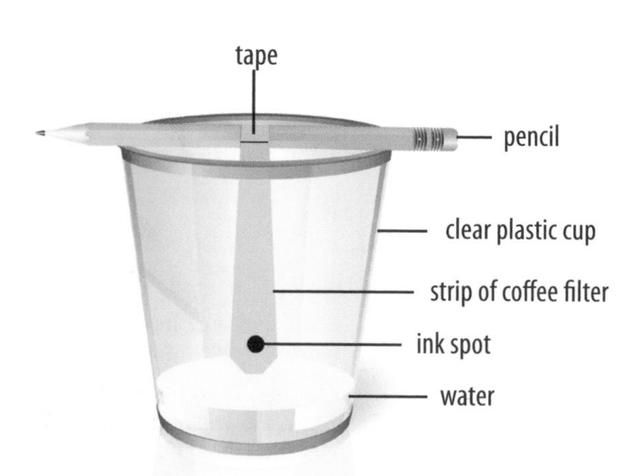

The bottom edge of the filter should be in the water, but the ink should stay above the level of the water. Hang all strips and observe for several minutes (or until the liquids have climbed about ¾ of the way up the filter). Remove strips from the water and blot on paper towel. The colored patterns that were produced are called chromatographs. Describe what you observed the colors doing. Record results.

The original colors are listed under "Colors tested." Record what happens to each color or set of colors under "Results." As best you can, name the separate colors you see on each chromatograph strip.

Colors Tested	Results (Colors seen after test)
Black	
Red & Blue	
Blue & Yellow	
Red, Blue & Yellow	

The Science Stuff

Some inks or colored markers are made up of a mixture of different chemicals. These mixtures can be separated without changing the chemicals. Chromatography is one way colored chemicals in a mixture can be separated and observed.

Chromatography

Water is able to dissolve the dyes in the colored markers. The water can move up the paper by a process called capillary action. It is much like how a towel will get wet if just the corner is left in a bathtub. The water carries the dissolved dyes up the paper with it, but some of the dyes may be carried faster than others.

If there is a strong attraction between the dye and the filter paper, the dye will travel slower; if there is little attraction, the dye will travel faster. Some dye molecules are heavier than others, and this will also affect how fast they are carried up the paper.

After several minutes, there will be a separation of the dyes as some of them are carried up the paper by the water faster than others. Chromatography can also be used to separate mixtures that do not contain dyes. They can even separate and identify colorless substances if they are stained later.

The dyes in the colored markers underwent physical changes when they were combined with each other, when they were dissolved in water, and when they moved up the filter paper. Each dye kept its own properties during this activity.

Dig Deeper

Do this same activity using food coloring. Apply the food coloring with a cotton swab or try this same activity using different markers, or try this same activity using a different kind of liquid. (Be sure it is a safe liquid.)

Making Connections

Forensic scientists can use chromatography to identify specific kinds of dyes. For example, a particular kind of ink can be identified if a test ink produces the same pattern. Each colored spot represents at least one chemical. Specific kinds of dyes can often be identified from this method.

What Did You Learn?

1. Is paper chromatography used to separate mixtures or compounds?
2. Were some of the dyes carried up the paper moving faster than others?
3. Do the chemicals in a mixture keep their own properties?
4. If two samples of ink produce the same chromatography pattern and colors, what would this indicate?
5. Suppose a chromatograph was made from a colored marker, and the pattern showed a blue spot above a pink spot. Does this give you a good idea that there are at least two chemicals in the colored marker?

INVESTIGATION #5

How in the World Can You Separate a Mixture of Sand and Salt?

Think about This

Observe the following demonstration. Your teacher will mix iron filings and sand together. We know this is a mixture, because the iron and the sand do not combine chemically with each other and they both retain their own physical properties. Iron and sand have different physical properties. See if you can predict what physical property your teacher will use to separate iron and sand. Were you right?

The Investigative Problems

How can you separate a mixture of sand and salt on the basis of one or more physical properties?

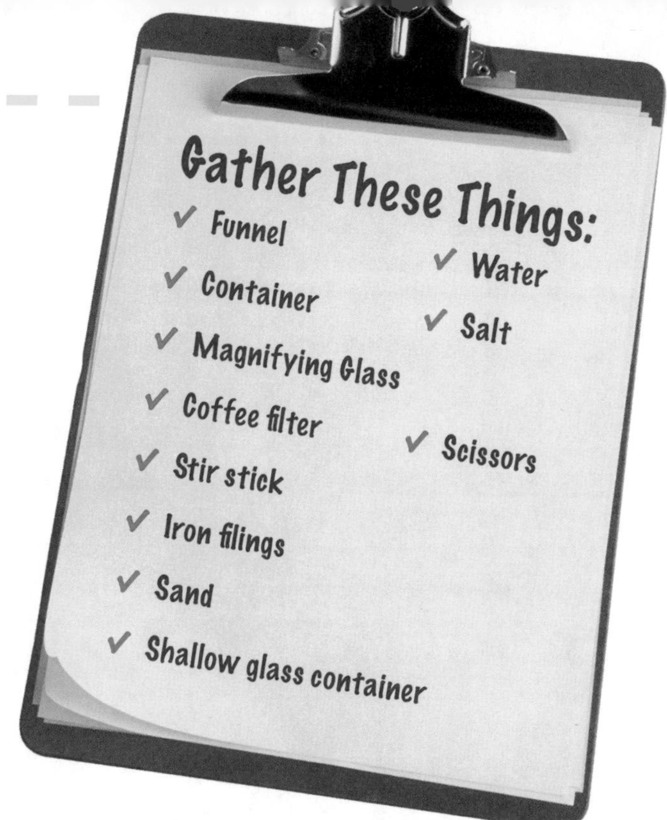

Procedure & Observations

1. Combine several spoonfuls of sand and salt. Stir them together. Observe the mixture with a hand lens. Describe your observations.
2. Place a mixture of salt and sand in a container and add enough water to dissolve the salt. Stir well. You can no longer see the salt, because the salt dissolves (is soluble) in the water. The sand will still be visible, because sand will not dissolve in water. What are at least two properties that sand and salt have that are different? →
3. Pour the contents of the container through a funnel that is lined with filter paper.
4. Be sure the sand is raked out of the container. Can you see anything other than a clear liquid in the substances that passed through the filter paper?
5. Lay the filter paper and sand flat on the table to dry. When it is dry, examine it with a hand lens. Does the dry sand look like it did when you observed it before?
6. Take a small amount of the filtered liquid and pour it into a shallow glass container. Fan the liquid for a few minutes. What do you see in the dish when the liquid evaporates? Examine the recrystallized salt with a hand lens. (The salt crystals may differ in size from the original crystals, because crystal formation depends on several factors. It is still salt.)

How to fold coffee filter

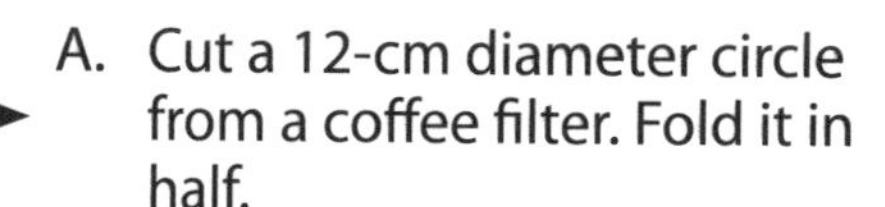

A. Cut a 12-cm diameter circle from a coffee filter. Fold it in half.

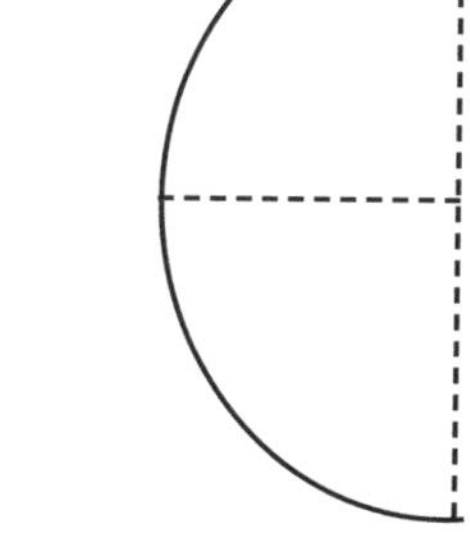

B. Fold it in half again.

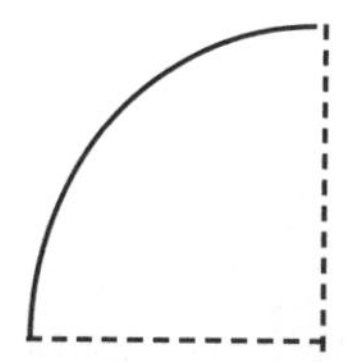

C. Open the paper to make a cone shape with three thicknesses on one side and one thickness on the other.

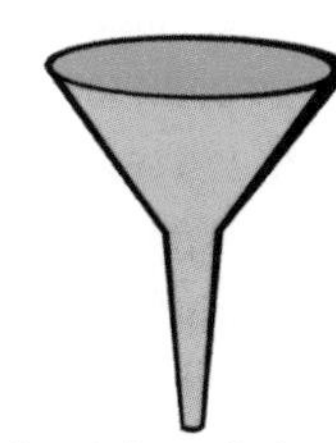

D. Wet the inside of the funnel and fit the cone in the funnel. Press firmly on the paper.

The Science Stuff

Sand is a pure substance; salt is a pure substance; and water is a pure substance. Every piece of a pure substance has the same properties as the whole substance.

Sand and salt have different physical properties. Sand and salt stirred together is a mixture of two substances. When water is added to the sand and salt mixture, nothing reacts chemically. Therefore, the combined sand, salt, and water is also a mixture.

Each substance in a mixture retains its physical properties. The salt dissolves in water and the sand doesn't. Water evaporates from the wet sand, but the sand doesn't evaporate. In a salt solution, only the water evaporates; the salt remains and reforms salt crystals.

You observed several physical changes in this activity. Mixing sand and salt was a physical change. Dissolving salt in water was another physical change. Evaporating the water from the wet sand and salt water were also physical changes. In a physical change, substances change in some way, but they keep their physical properties.

When two or more pure substances are mixed together, they keep their own properties. When two or more pure substances are combined chemically, they do not keep their own properties. They undergo a chemical change and acquire new properties.

Making Connections

Distillation is a process of separating chemicals on the basis of their boiling points. Gasoline for our cars, as well as many other products, comes from the process of distillation. Crude oil comes from underground reservoirs as a mixture of thousands of chemicals. This means that each chemical in the mixture has a different set of physical properties. The physical property most often used as a way to separate the chemicals is their boiling points. The process is known as **fractional distillation** and is done in large refineries.

In the process of fractional distillation, just enough heat is applied to cause the chemicals with the lowest boiling point to boil first. These chemicals are collected, cooled, and separated from the rest of the oil. Then the temperature is raised some more until it causes the next group of chemicals to boil and be separated. This process is repeated many times until eventually all the chemicals in the crude oil mixture are separated.

Dig Deeper

You may have seen bottles of water in a grocery store labeled "distilled water." Try to find out what this means and how distilled water is produced.

Make a diagram to show how an oil refinery works. Show where the crude oil is heated and turned into a gas where it is cooled and turned into liquid, and where the different groups of chemicals are separated. List the names of the major groups of chemicals that are produced in a refinery.

Most crystals grow slowly over several days or weeks. Try growing some crystals of Epsom salts, which can be observed in seconds or minutes. Use a medicine dropper and put four full droppers of water in a small container. Add a little Epsom salt until some of it will no longer dissolve in the water. Pour off most of the clear solution into a second container. Add a drop of Elmer's glue and stir well. Put several drops on a glass slide and wait a few minutes. Observe with a magnifying lens as the liquid dries. Draw some of the crystals. Explain under what conditions crystals usually form.

Try to "grow" some alum crystals and some sugar crystals. Search the Internet or books from the library for a variety of methods that can be used to grow crystals and choose one that looks interesting. Compare the ways in which they are alike and the ways in which they are different. Make pictures of them or draw the shapes of the crystals.

What Did You Learn?

1. Which of the following are examples of mixtures: salt and sand stirred together, crude oil, salt water, distilled water?
2. What is one way a mixture is different from a pure substance?
3. When two or more pure substances are mixed together, do they keep their individual properties?
4. When two or more pure substances are combined chemically, do they keep their individual properties?
5. Distillation is a way of separating mixtures of liquids. This process depends on differences in what physical property?
6. What kind of substance can be separated from a liquid by a funnel and filter paper — one that is dissolved in the liquid or one that is not dissolved in the liquid?
7. How can you separate a mixture of salt and water?

Learning about Water

We will spend the next few lessons learning what an incredible substance water is. We will also look at water solutions of acids, bases, and salts. God has designed water as the most special substance to sustain life on this earth.

Water was one of the first things God made during the week of creation because all living things need water. How much do you know about references to water in the Bible?

Quiz

Let's start with a little quiz about the word *water*.

1. Do you know the first time the word water is mentioned in the Bible?
2. What did God do with the waters on the second day of creation?
3. What did God do with the waters on the third day?
4. What did God do to the water on the fifth day?
5. To whom did God ask this question: "Who shut up the sea behind doors when it burst forth from the womb. . . when I fixed limits for it and set its doors and bars in place"?
6. What great writer in the Bible makes the observation that all the rivers run into the ocean, but the ocean never gets full?
7. To whom did Jesus tell," If you knew the gift of God and who it is that asks you for a drink, you would have asked him and he would have given you living water"?
8. Who said, "Whoever drinks the water I give him will never thirst"?
9. Peter wrote (2 Peter 3:6) that "the world that was" perished. How did this happen?

Water Is the Standard

Think about This

Work with a partner. You will both be given an object, but don't look at each other's objects. Observe your object and try to describe its size and shape using words and numbers only. Use no more than 20 words/numbers. See if your partner can draw an accurate picture of your object based on your description.

The enormous energy of Niagara Falls has long been recognized as a potential source of power.

Procedure & Observations

Show your partner the object you described. How accurate was your partner's drawing? Look at your drawing. How accurately did you draw your partner's object? An important part of being a scientist is the ability to provide accurate descriptions and measurements.

Try to put 30 mL of water into your graduated cylinder (also called a graduate). This may not be as easy as it sounds. When you add water, it will look like the surface of the water is curved and in two places. Be sure the graduate is sitting on a flat, level surface and your eyes are even with the water's surface. To correctly tell how much water is in the cylinder, look at the bottom of the curve that forms on the surface of the water. Read the number that is closest to the bottom of the curve. This is the volume of the water in the cylinder, minus a few drops. You should record the volume of the water in milliliters.

Tie a string around the rock and gently lower it into the cylinder. Determine the water level again. If you subtract 30 mL from this water level, you can determine the volume of the rock. This method is known as the displacement of water and is often used to find the volume of an irregular solid. What is the volume of the rock?

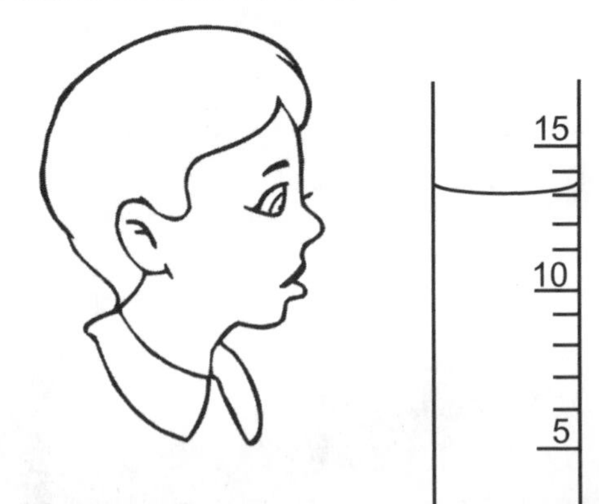

The correct reading is 13mL. You read the bottom of the curved surface of the water and other liquids.

The Investigative Problems

How do scientists find volume, mass, and density?

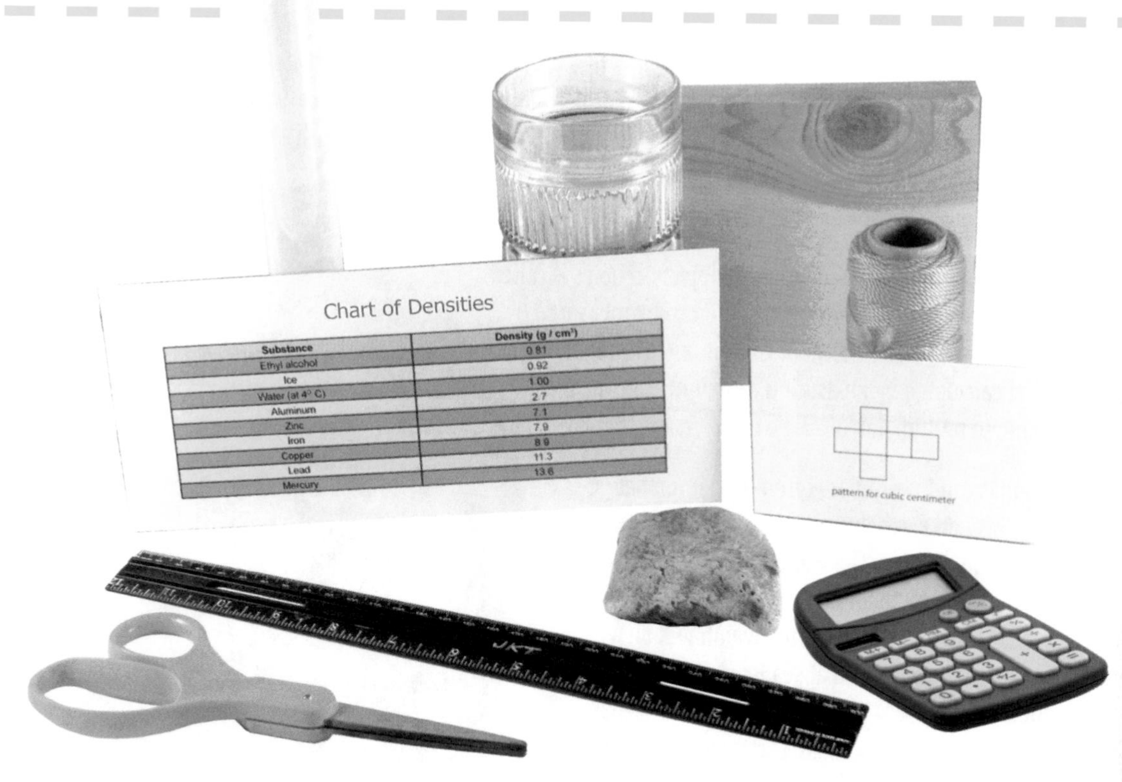

You can find the volume of a regular solid by measuring the sides and doing some calculations. You could measure the rectangular wooden block using inches or centimeters. Most scientific measurements are done in metric units, so we will use the metric ruler.

But first, look at the ruler that is marked off in inches. Halfway between an inch, there is a midpoint mark. Halfway between this distance, there is another midpoint mark. Each of these distances may be divided again, so each inch will be marked off into eight sections. (Some rulers mark each inch into 16 or even 32 sections.) Draw a line that is 3 ⅝ inches long. Start marking your line at the first long line and continue drawing your line to the long line by the number three. There will be shorter lines after the three where the inch is divided into ⅛-inch sections. Count five of the shorter lines past the number three and end your line there. If you had a block that was 3 ⅝ inches on each side, you would multiply 3 ⅝ inches times 3 ⅝ inches times 3 ⅝ inches to calculate the volume of the box in cubic inches. What is the volume of the block? (Convert 3 ⅝ inches into 29/8 inches. You may use a calculator to help you multiply and divide.)

Now look at the metric ruler. This ruler is divided into centimeters (cm). Each centimeter is divided into ten sections. Draw a line that is 3.7 cm long. Start marking your line at the first long line and continue to the number three plus the next seven small lines.

Now measure and record the length, height, and width of the rectangular block you are given in cm. Multiply the three numbers to find the volume in cm³. What is the volume of the wooden block? (You may use a calculator to help you multiply.)

Most people find it easier to work with decimals rather than with fractions. In the metric system, everything is based on 10s. You can multiply or divide by 10 by moving the decimal right or left. For example, 5.5 times 10 = 55, and 5.5 divided by 10 = 0.55.

Cut out the pattern to make a cubic centimeter box (see appendix). If this box could be filled with water, the water inside the box would weigh one gram. If the water in the box were poured into a cylinder, it would measure one milliliter.

Use the density chart in the appendix to find the density of some common substances. If the box were filled with iron, what would it weigh? If it were filled with a type of wood, what would it weigh? If it were filled with ethyl alcohol, what would it weigh? Find three more substances on the density chart. Tell how much each would weigh if they exactly filled your cubic centimeter box.

The Science Stuff

Scientists have learned that they must work with specific and exact information. References to a "tall tree" or a "big lake" have no value in science unless measurements are taken and other information about the tree and the lake are included. There are many reasons why scientists try to describe things in such a way that someone else will know exactly what they mean.

A basic principle of science is that all matter takes up space and has **mass (weight)**. The amount of space an object occupies is called its volume. The volume of an irregular solid can be measured by the displacement of water method. The volume of a regular solid can be calculated by measuring its height, length, and width and then multiplying all three numbers.

The terms mass and weight are often used interchangeably at sea level on earth, but there is a difference in what they mean. Think of the weight of an object as a force that is caused by the pull of gravity on it. Weight can change. For example, you would weigh less on a mountaintop than you would at sea level. Your weight on the moon would be ⅙ what it is on earth. Your mass, however, would stay the same no matter where you were.

Water is such a common substance on earth that metric units use the mass and volume of water as a standard measurement. Every milliliter of pure water has a mass of exactly one gram. It isn't necessary to weigh a container of water to know what the water inside weighs. You can just measure its volume in milliliters and that is its weight in grams. This means that the 30 ml of water in your graduate weights 30 grams. (This only works for water.)

Another neat exchange of units is one milliliter = one cubic centimeter. So, you can also say that the 30 milliliters of water in your graduate has a volume of 30 cubic centimeters, and weighs 30 grams.

One of the most important physical properties of a pure substance (an element or a compound) is its density. The density of any substance can be calculated by dividing its mass by its volume. You can divide the mass of one gram by the volume of one milliliter to calculate the density of water. If you prefer, divide the mass of one gram by a volume of one cubic centimeter. Your answer will be one gram per milliliter or one gram per cubic centimeter. Either answer is the correct density for water.

In chemical shorthand, the density of water can be written as one g/mL or one g/cm^3 or one g/cc. Many doctors and scientists abbreviate cubic centimeter as cc simply because it is easier to type.

Since the density of water is one gram per cubic centimeter (g/cm^3), the density of any substance can be compared to water. Any substance that doesn't dissolve in water and has a density less than this will float on water; if its density is greater, it will sink. For example, the density of iron is 7.8 g/cm^3. Since its density is greater than one g/cm^3, a solid block of iron will sink in water, regardless of its size. If the density of a substance like wood is less than one g/cm^3, a solid block of any size will float on water.

Dig Deeper

When were standard international (SI) units of measurement first devised? What is used as a standard today to ensure that everyone knows the exact length of a meter?

Make a chart of metric units used to measure volume, length, mass, force, and temperature. Illustrate some ways that small units can be converted into big units and some ways in which big units can be converted into small units.

Convert the measurements of Noah's ark from cubits to feet or meters.

Fill a bathtub, an aquarium, or a large glass container about ⅔ full of water. Add a sealed can of a regular soft drink and a sealed can of the diet version of the same soft drink to the water. See if you can tell which one has the greater density by which one floats. Add other sealed cans of drinks to the water, but make sure each pair of cans is equal in volume and is made of aluminum. Review what you learned about calculating density.

Making Connections

For many years, people measured buildings and other objects by body parts. The distance of a man's foot was used to measure distance. The distance from a man's elbow to the tip of his middle finger was called a cubit. The dimensions of Noah's ark were given in cubits. When precise measurements are needed, standard measuring rulers are needed. Not everyone has the same size arm.

Most American tools are based on the English system of numbers instead of the metric system. Many countries in the world use the metric system. This can be a problem if someone owns a set of tools based on the English system and they purchase a vehicle that was built according to metric unit. For example, they may find that a ⅜-inch wrench is too small, but a 4/8-inch wrench is too big to turn a bolt.

What Did You Learn?

1. How would you find the volume of a small, irregularly shaped piece of metal?
2. What happens to a person's mass as the distance from the earth increases? What happens to the person's weight?
3. Give an example of a metric unit that is used to measure volume.
4. Give an example of a metric unit that is used to measure how long something is.
5. Give an example of a metric unit that is used to measure an object's mass.
6. What is the density of pure water? Write this as words and as symbols.
7. If a substance has a density of 2.5 g/mL, will a block of this substance float or sink in water?
8. How can you calculate the density of a substance?
9. How much does 25 mL of water weigh?
10. How can you find the volume of a large rectangular wooden block?
11. The displacement of water method is used to find the volume of a rock. Its volume is found to be 19 mL of water. What is the volume of the rock in cm^3?

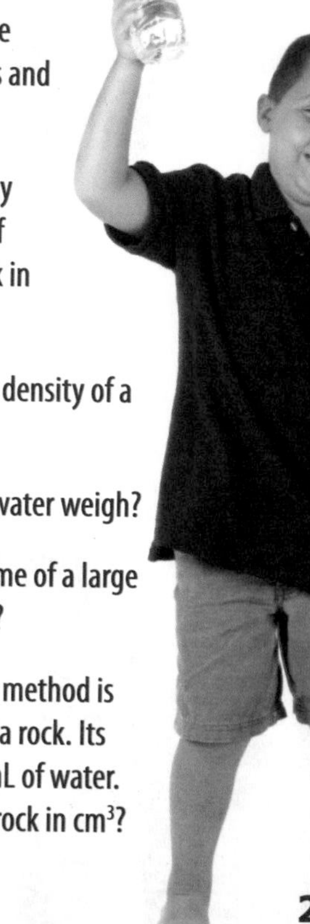

Bending Streams of Water

Think about This

There is no written record of ancient, scientifically minded people doing the experiment in today's lesson. However, they had the means to do this as far back as 600 B.C. The Greeks discovered that a substance called amber (a plant resin) could be rubbed with wool and it could then attract bits of fluff. Who knows, someone may have figured out that a piece of charged amber could cause a thin stream of water to bend. But it would take over 2,300 years for anyone to start to understand why rubbing amber with wool would cause it to have an effect on some other substances.

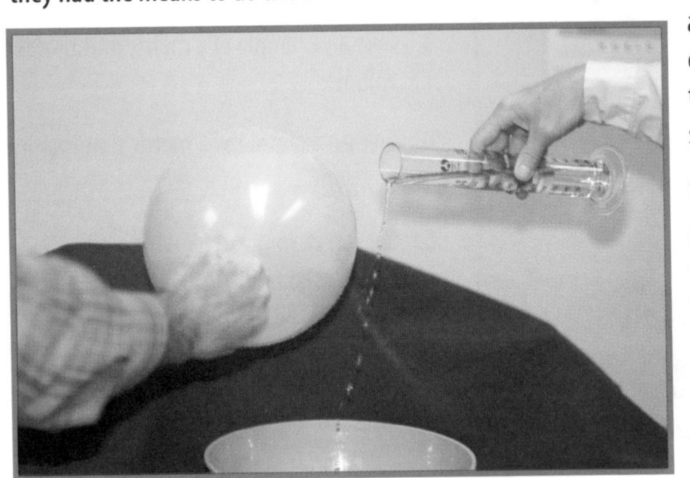

We still haven't been able to actually observe some of the world's tiniest particles as they move about. However, you have the benefit of learning the very logical explanations of many brilliant thinkers and experimenters. The discovery of positive and negative particles is at the heart of the field of chemistry.

The Investigative Problems

Are there positively and negatively charged particles in matter? Can a charged object attract a stream of water? What kind of shape do water molecules have?

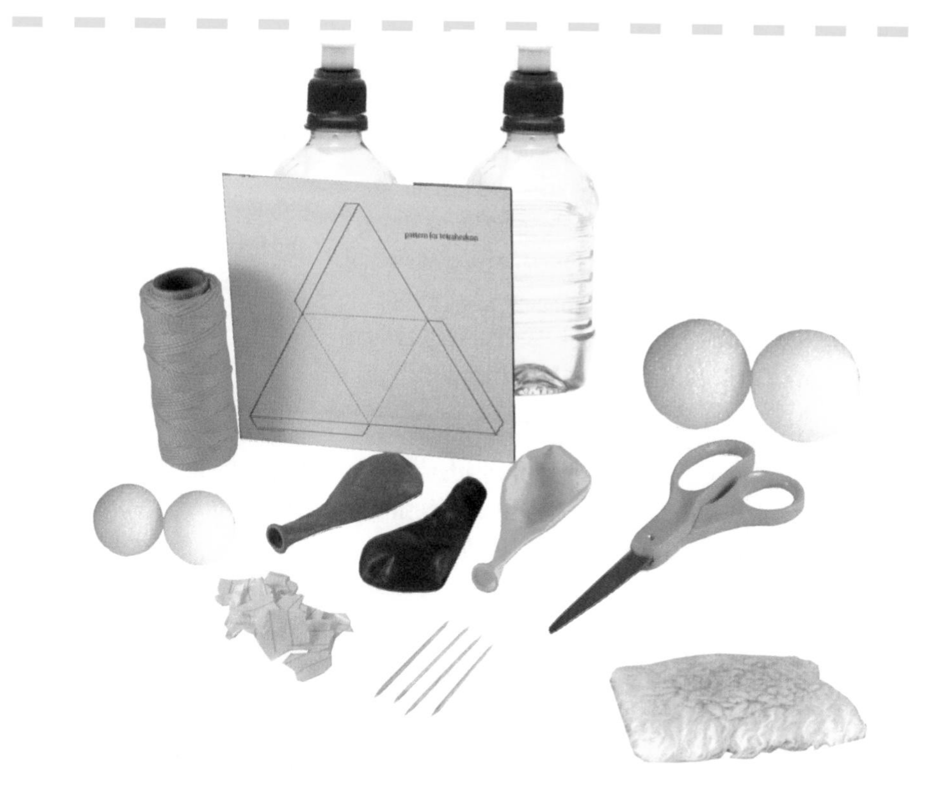

Procedure & Observations

Charge a rubber balloon (or comb) by rubbing it with the wool. Move the balloon over small pieces of paper without touching the paper. Do you see some of the paper moving? Record what you see. Touch the balloon firmly with your hand in several places. Move the balloon over the paper again. Does anything happen this time?

Charge the balloon again by rubbing it with the wool. Make a small, steady stream of water and hold the balloon close to it. Don't let the balloon touch the water. Observe how the stream of water is affected by the balloon. Record your observations. Draw a picture of what you observed with the balloon and the stream of water.

Cut out the pattern for the tetrahedron. Fold the paper to make a tetrahedron and put the 1-inch Styrofoam ball inside of it. Take four toothpicks and stick a toothpick in each corner of the tetrahedron, pushing each toward the center of the Styrofoam ball. Remove the paper. Each toothpick should be just as far away from one as it is another. When you get your toothpicks in place, turn the ball over several times. It should look the same each time you turn it over. Are any two toothpicks on opposite sides of the ball?

Stick two small Styrofoam balls into any two toothpicks and observe the arrangement. Now remove the balls and stick them to the other toothpicks. Rotate the balls until this arrangement looks the same as the first one. This is a model of a molecule of water. If you have assembled everything correctly, it will look much like a teddy bear face. Draw a picture of your water model with 2 hydrogen atoms and 1 oxygen atom and label it "model of a water molecule."

Remove the toothpicks and reinsert them in the large Styrofoam ball, so that two of them are on opposite sides of the ball. Stick two small Styrofoam balls into these toothpicks. If the two hydrogen atoms were arranged like this, water would have very different properties. Do you see the difference between this arrangement and the one above? Draw a picture of this arrangement and label it "not a model of a water molecule."

Describe at least four physical properties of water that you can observe with your senses. (Consider things such as color, odor, texture, state, etc.) Water has many other physical properties that can be observed using thermometers and other equipment. You will learn in the following lessons that many of the physical and chemical properties of matter are the result of the arrangement of the positive and negative charges of which they are made.

The Science Stuff

All matter is made of positively and negatively charged particles, as well as neutral particles. We don't usually notice these charges since they tend to be evenly balanced most of the time. The two examples of **static electricity**—a charged balloon attracting small pieces of paper and a charged balloon causing a stream of water to bend—showed times when positive and negative charges were not balanced. We have all observed other examples where charges are not balanced, such as lightning, clothes with static cling, or current electricity.

Scientist have known for many years that positive and negative charges attract each other, regardless of whether they are touching or not. Furthermore, two like charges (positive and positive or negative and negative) repel each other. The attraction or repulsion becomes stronger as charges get closer together.

Charges that result from static electricity don't usually last long. Did you notice that the balloon lost its charge when you touched it with your hand?

However, even when there is no static electricity, water molecules have a permanent slight imbalance of charges. The shape of the molecule is part of the reason for this.

A water molecule looks somewhat like a teddy bear face. The small balls in your water model represent hydrogen atoms; the large ball represents an oxygen atom. Each toothpick represents a pair of negatively charged particles called **electrons**. The end of the oxygen atom with the four negative electrons will be slightly negative and the end that is attached to the two hydrogen atoms will be slightly positive.

Strong chemical forces or bonds hold the oxygen and hydrogen atoms of water together. These bonds are strong enough that most of the water molecules will not break apart, even if water is boiling or freezing. Because water molecules usually hold together and have a slightly positive end and a slightly negative end, they are said to be **polar molecules.**

If the two hydrogen atoms were located on opposite sides of the oxygen molecule, the molecule would not be polar. Seeing the geometry of a water molecule can be helpful in understanding why this is true. The four corners of a tetrahedron represent the possible places where two hydrogen atoms can join an oxygen atom to make a molecule of water. No matter where two hydrogen atoms are attached, they will not be opposite each other.

This is a good time to learn the meaning of the terms **atoms** and **molecules** as they relate to water, hydrogen, and oxygen. A water molecule is made up of two atoms of hydrogen and one atom of oxygen. These three atoms are tightly bound together by chemical forces.

The **chemical formula** for water is H_2O, again showing that a water molecule is made up of two atoms of hydrogen and one atom of oxygen. Although water molecules are too small to observe, there are many reasons to believe that scientists have correctly figured out how water molecules are arranged.

Hydrogen gas only contains hydrogen atoms, and oxygen gas only contains oxygen atoms. If hydrogen gas and oxygen gas are mixed together and someone throws a lighted match into the mix, there will be an explosion. Some of the hydrogen and oxygen atoms will join together and form drops of water all around the container. Water has very different properties from either hydrogen or oxygen.

Water is a remarkable substance with an interesting combination of properties. It forms round droplets, it expands when it freezes so that ice floats on water. Unlike other small molecules, water is a liquid at room temperature. Water cannot dissolve things like fats and oils, sand, and metals. However, it is able to dissolve many other chemicals. Almost all salts, acids, bases, sugars, and many other things are **soluble** (can dissolve) in water.

Many of the properties of water are the result of a slight imbalance of charges that remain in the water molecules, causing them to be polar molecules.

Making Connections

Scientists have been very creative in figuring out that atoms contain positive, negative, and neutral charges that they combine to form molecules and that molecules have specific shapes. They have been able to do this without directly observing any of these things. Throughout the next several lessons, we will look at how a few of these things were figured out.

For hundreds of years, scientists thought water was one of the earth's simplest elements. It could be heated, boiled, or frozen and it still remained water. It seemed logical to conclude that water couldn't be broken down into something simpler. Still, the early scientists were mistaken. Water can be broken down into hydrogen and oxygen under some conditions.

Many "spoofs" are found on the Internet that are purposefully misleading to be a joke. One spoof concerns the chemical "dihydrogen monoxide" (DHMO). This chemical is portrayed as a harmful chemical that can be fatal when inhaled, is a major component of acid rain, and withdrawal can result in violence. It is suggested that DHMO should probably be banned.

See if you can figure out the one-word common name for this chemical. The reason this spoof is so believable is that the statements are accurate facts. Ask your teacher if you need help figuring out the common name of this chemical.

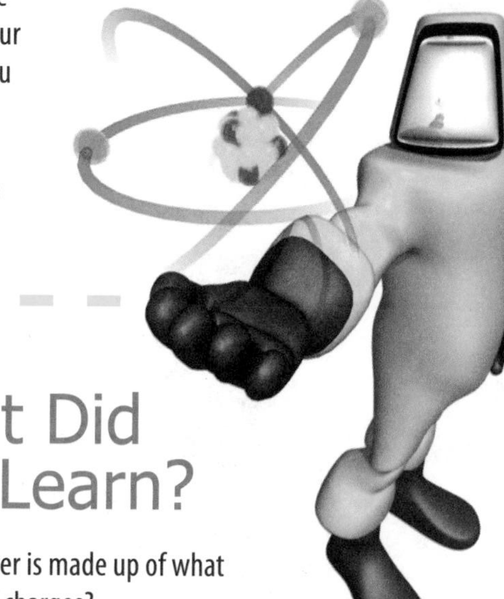

Dig Deeper

When water freezes and becomes a solid, it floats instead of sinking. Water is heaviest at about 4°C, so cold water sinks to the bottom of a pond, but it doesn't freeze. Try to find out what happens to the fish and other cold-blooded animals that live all winter in a frozen pond. What would happen to these animals if the pond started to freeze at the bottom, which is the way most other liquids freeze?

What is meant by "water table"? Sometimes large areas, or even entire states, go through a period of drought when there isn't enough rain to keep the water table up. Find an area where there has been a recent drought. What happens to the water table? What problems are caused when the water table in the area changes as a result of a drought? Find an area where there has been recent flooding. What happens to the water table when there is severe flooding of an area?

Water forms weak hydrogen bonds that link one water molecule to another. Try to find out more about how hydrogen bonds form. Draw pictures to illustrate this. Try to find out how hydrogen bonds affect water's boiling point, freezing point, and liquid state.

What Did You Learn?

1. All matter is made up of what kinds of charges?
2. Name some things that can be dissolved in water.
3. A water molecule is made up of which two kinds of atoms?
4. What geometric shape explains one reason why water molecules are polar?
5. What are the molecules called that have strong connecting bonds, a positive charge on one end, and a negative charge on the other end?
6. List at least four physical properties of water.
7. Write the formula for water and explain each symbol and number.
8. What is the difference between an atom and a molecule?

Drops of Water

Think about This

Pour ¼ cup of water into a graduated cylinder on your table. See if you can read the amount of water in the cylinder in milliliters. Remember, the surface of the water will have a curved shape and it will look like there are two water levels. Think about why water drops take on round shapes, but when water is inside a graduated cylinder, it curves the other way.

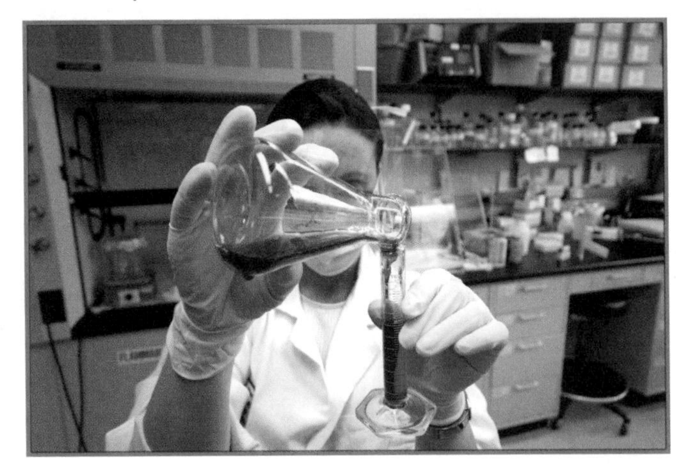

The Investigative Problems

Why is the shape of a drop of water round? What is the shape of the surface of water inside a graduated cylinder? How can you change the shape of a drop of water?

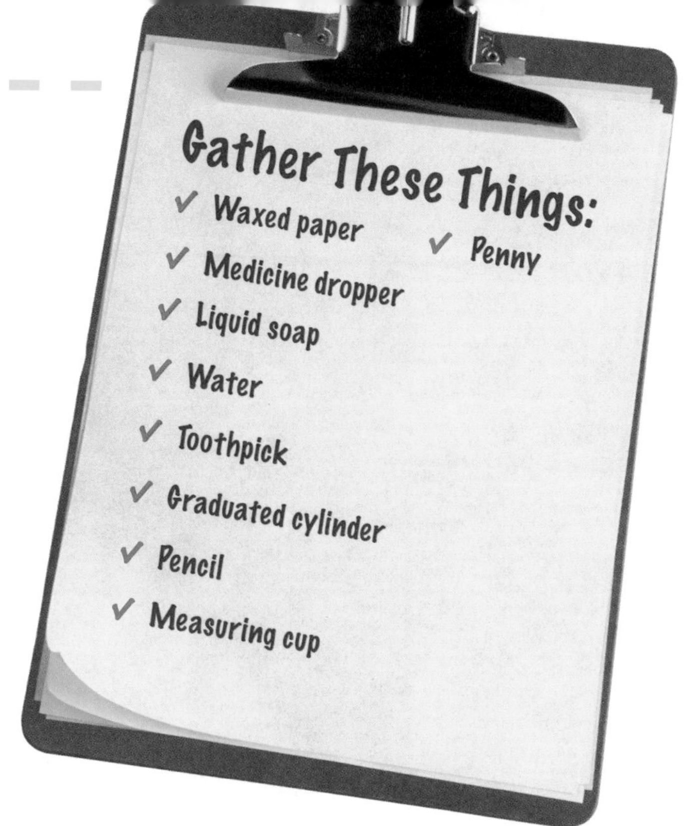

Procedure & Observations

Hold the dropper about one cm above the waxed paper. With the same amount of water in the dropper each time, place five drops of water onto the waxed paper, keeping each one separate from the others. How do they compare in size?

Put the sharpened point of your pencil into one of the drops of water and carefully observe how the water responds when the pencil is lifted out.

Touch the eraser of your pencil to another drop of water and carefully observe how the water responds when the eraser is lifted out.

Are the water molecules more attracted to the lead end or the rubber end?

With the point of the pencil, push a third drop around the paper. Push two drops together, then three and four. Pay attention to the details and write your observations.

Place three or four more drops of water on the wax paper with the dropper. Dip the end of a toothpick in a little liquid soap and touch each drop of water with the toothpick. What happens to the drops of water when soap is added?

Predict how many total drops of water you can place on a penny before it overflows.

Now test your prediction by putting one drop of water at a time on a clean, dry penny with a medicine dropper. Keep the dropper close to the water on the penny, but don't touch it with the dropper. Count the number of drops you are able to add before some of the water overflows. Record the number of drops that will sit on a penny.

Repeat the experiment, but this time use one of the following solutions — salt water, soapy water, or rubbing alcohol and water. Count the number of drops of the second solution you are able to add before some of the solution overflows.

The Science Stuff

Remember that water molecules do not break apart easily. They are **polar** because one end is slightly positive and one end is slightly negative. The positive end of one water molecule can attract the negative end of another water molecule.

The following diagram explains why molecules of water are attracted to each other.

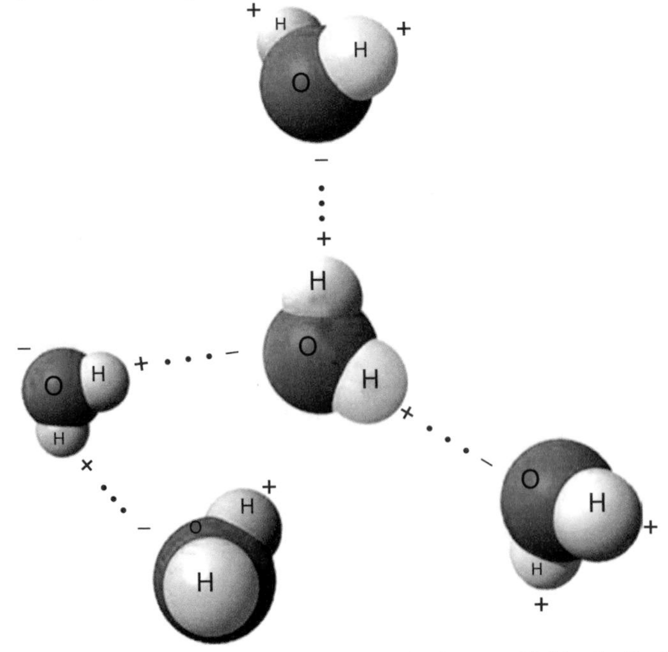

The attraction of similar molecules for each other is called cohesion. Within the liquid this attraction between molecules occurs equally in every direction. In the top layer of molecules, however, they are pulled downward and sideways but not up. This unbalanced attraction of the top layer of water molecules is called surface tension.

The roundness of the drops on the waxed paper is an indication of surface tension. When several drops are put together they flatten out more because of the increased effect of gravity. If there were no gravity, a drop of water would take on the shape of a perfect sphere.

The round water drops can also be flattened when a little soap is added, because soap breaks down the surface tension of the water drops.

The attraction that two different kinds of molecules have for each other (as glass for water) is called adhesion. A few drops of water stick to the glass or plastic sides of a graduated cylinder because of adhesive forces. The adhesive forces between the water and the container are stronger than the cohesive forces between water molecules. This is why the water inside the graduate doesn't have a rounded shape like the drops on the waxed paper.

A surprising number of drops of water can be placed on a clean penny before it runs off. The water on the penny will form a rounded shape because of the strong cohesive forces between the water molecules, especially at the surface of the water.

Making Connections

The surface tension of water forms a bond strong enough to float small pieces of metal laid carefully on the water. When detergent is added to the water, the surface tension is broken and the "floating" object will sink.

Many other substances have surface tension. It has been suggested that better ball bearings could be formed in space than in factories on earth because a drop of molten steel in space would naturally form a perfect sphere.

Dig Deeper

See if you can gently lay a small piece of metal, such as a sewing needle or the spring out of a ballpoint pen, on the surface of the water in a container. You may need to lower it onto the water with a couple of thin pieces of string. After you get it to lie on top of the water, add a drop of liquid detergent (or put the edge of a bar of soap) to the water and observe what happens. Give an explanation for this.

Place several drops of water on a piece of wax paper with a medicine dropper. Touch the water drops with a toothpick that has been dipped in a substance such as rubbing alcohol. Test several different substances to see which ones affect the surface tension of water.

Some kinds of insects "skate" on top of water, because they don't break through the water's surface tension. Find out more about insects that "walk on water."

What Did You Learn?

1. One end of a water molecule is positive and one end is negative. What happens when the positive end of one molecule comes near the negative end of another molecule? Make drawings to illustrate this.
2. What is a polar molecule?
3. At the surface of the water, the molecules attract each other in every direction except up. What name is given to the attraction of water molecules at the surface?
4. Why do water molecules have an attraction for each other?
5. Adding soap or detergent to a drop of water causes the drop to flatten out. What does soap break down when it is added?
6. To correctly read the amount of water in a graduated cylinder, what part of the curved shape of the surface of the water must one read? Make a diagram if you wish.
7. Is there an attraction between the water and the glass (or plastic) in a graduated cylinder? Is this attraction known as cohesion or adhesion?
8. If you divide the mass of a substance by its volume, what would you calculate?
9. If several drops of water are placed on a clean penny, what kind of shape will the water have? Why is this?

Oil and Water Don't Mix

Think about This

You have probably noticed that it is difficult to get greasy hands clean if you just wash them with water. Rub a fatty substance or oil on both hands. Now try to wash your hands using water only. Think about how you can get the fatty substance off your hands.

The Investigative Problems

Is there a way to make oil and water mix?

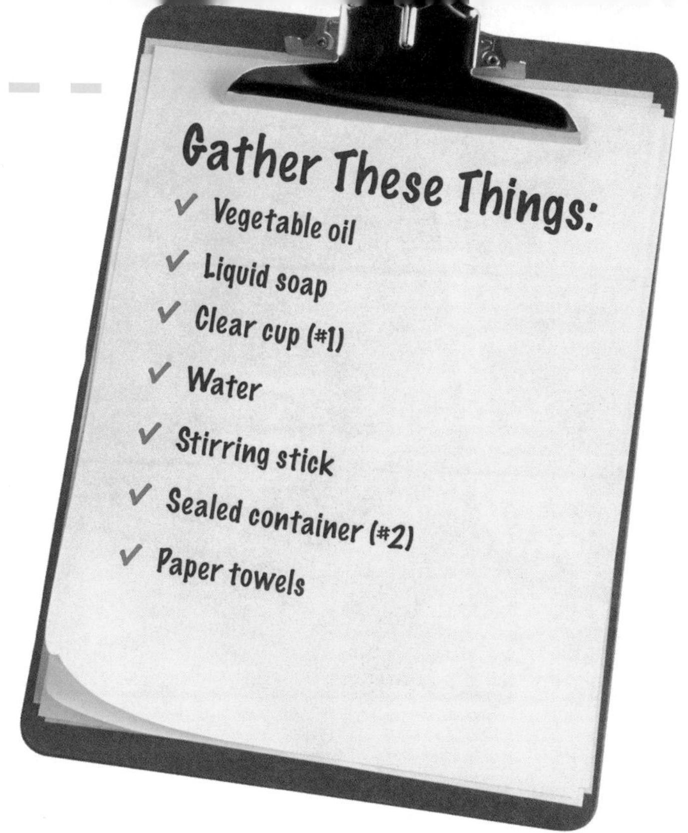

Procedure & Observations

Pour equal amounts of oil and water into the cup (#1). Write your observation. Stir the mixture with a stick and make further observations.

Pour equal amounts of oil and water into the container (#2), close the lid, shake several times, and observe. Write your observations.

Add a few of drops of soap to the cup (#1) and stir well. Did the oil and water mix in the cup when stirred?

Add a few drops of soap to the container (#2) with the mixture of oil and water. Close the lid and shake the container several times. What did the oil and water do this time?

What difference did the addition of a few drops of soap make to the mix?

Now rub a fatty substance on your hands as you did before. Wash your hands with soap and water. Did the fatty substance come off easily this time?

The Science Stuff

You have seen that oil and water don't mix. The reason is that water molecules are polar. Oil molecules have evenly balanced charges and are **non-polar molecules**.

Polar molecules will dissolve other polar molecules, as well as some other kinds of particles with positive and negative charges. Non-polar molecules will dissolve other non-polar molecules. Non-polar molecules (like oil) don't dissolve in polar molecules (like water).

You have also seen that oil and water will mix when soap is added. Soap molecules are long structures that are polar on one end and non-polar on the other end. The polar end of the soap dissolves the polar water molecules. The non-polar end of the soap dissolves the non-polar oil molecules. This quality of the soap allows the two molecules of oil and water that don't want to get together to come together. Detergents and other cleaners work in similar ways, but often have features not found in soap.

Making Connections

For many years, homemade lye soap was the main source of soap for many farming families. It was combined with water and used for washing hands, hair, dishes, clothes, and anything else that needed cleaning.

Soap is a type of salt made by combining fatty acids with a base. It is similar to detergents and other kinds of cleaners. Today's cleaners come with many choices, such as scented, low-suds, softeners, emulsifiers, wetting agents, or biodegradable features.

Dig Deeper

Find an old piece of cotton fabric, and soak it in an oily substance. Add a little lipstick and oil-based makeup for color. When it dries, cut it into five-inch squares. Soak each square in a different kind of cleaner for several hours. Soak one square in water only. Be sure to use equal amounts of the cleaners. Rinse the squares without scrubbing the cloth to see which cleaner did the best job of getting the oily stains out. Make a display of your results and tell how each square was treated.

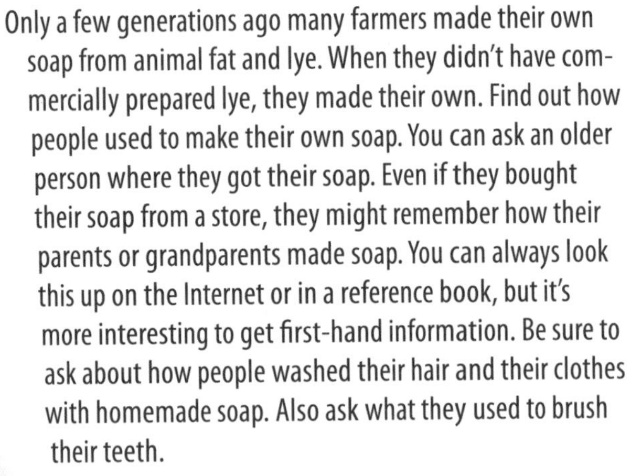

Only a few generations ago many farmers made their own soap from animal fat and lye. When they didn't have commercially prepared lye, they made their own. Find out how people used to make their own soap. You can ask an older person where they got their soap. Even if they bought their soap from a store, they might remember how their parents or grandparents made soap. You can always look this up on the Internet or in a reference book, but it's more interesting to get first-hand information. Be sure to ask about how people washed their hair and their clothes with homemade soap. Also ask what they used to brush their teeth.

What Did You Learn?

1. Some of the most important properties of water occur because water molecules are slightly positive at one end and slightly negative at the other end. What are these kinds of molecules called?
2. Are oil molecules polar or non-polar?
3. What can you generally predict about dissolving non-polar substances in polar substances?
4. How is soap able to dissolve both polar and non-polar substances?
5. Why is it hard to get oily substances, such as lipstick, out of clothing using just water?
6. Why is it hard to wash oil off your hands only using water?

NARRATIVE: Is There Life on Other Planets?

Joey read the headline: "We Are Not Alone in the Universe."

"Wow, Dad," he exclaimed as he bounded into his father's office holding up the paper. Dr. Houston, a chemistry professor at the university, looked up from reading a chemistry journal. "Isn't this exciting?"

"Have you read the entire article?" his father asked. "Tell me what you think has been discovered."

"It says some scientists found a new planet far out in space and there is life on it. Isn't that what it says?"

"Not quite," Dr. Houston said. He waited a few minutes to give Joey a chance to read past the headlines. "The evidence is quite scanty. Remember, scientists must always examine their evidence carefully before they come to a conclusion. This sounds like a newspaper reporter may have drawn a few personal conclusions based on imagination and excitement."

"But why would the paper print things that aren't true? Couldn't they get sued or something?" Joey was perplexed. The article sounded to him like scientists had discovered life on another planet.

"Let's look at the evidence, Joey. I read an article about this research in a scientific journal last week. The researchers designed a special piece of equipment that they attached to a large telescope. This device detects wobbles in certain waves that are given off by stars. What they found were slight wobbles in a star that could be caused by the gravity of an orbiting planet or planets.

"This is an interesting technique, which may give evidence about possible planets that orbit other stars," Dr. Houston continued. "The researchers believe there are three planets, but they are most interested in the second one. They have interpreted their data to mean that it is five times more massive than the earth and it orbits the star every 13 days, because this is a small star."

"You mean, no one has seen this planet, and they're not even sure it exists? Why do they think there is life on it?" Joey was really confused now.

"If their data is correct about its size and distance from the star, then there is a possibility that there could be liquid water on its surface. The presence of liquid water doesn't mean there is life, but water is necessary for life."

"Why do they think there is liquid water on it?" asked Joey.

"There is no evidence that water exists there. The scientists are looking at the possibilities that if water does exist, then it might be in liquid form. The planet is so close to the star that it would probably rotate with the same side facing the star at all times. That's how our moon rotates around the earth. We only see one side of the moon from earth. In the case of the planet, the star-facing side might be very hot and any water there would evaporate. The other side might be very cold and any water there would freeze.

"Nevertheless, there could be areas in between the hot and cold sides where liquid water exists," he continued. "It depends on the kind of materials that make up the planet and whether or not there is air around it. Another reason to suspect liquid water is that the planet is large enough to have a strong gravitational pull. This would hold most liquids and vapors to the planet."

"If liquid water was present, would this planet have other things that would be needed for life?"

"Even if water were present, it is highly unlikely that life could exist under these conditions."

"So," Joey began cautiously. "Scientists think there is a planet because of some wobbles. They think there might be liquid water in some areas on this planet. They think life might have evolved in the areas where there might be liquid water. Is that right?"

"That's the logic," Dr. Houston replied.

"Are there any other planets that might have all the right conditions for life to exist on them?" Joey saw in the article that scientists might have spotted as many as 200 likely planets.

"No, I'm afraid not. Some planets are too hot and all the water would evaporate. Other planets are too cold and all the water would be frozen. Some are too small and don't have enough gravitational pull to hold air or liquids on them. Some are too big, which would cause everything to weigh too much. There are many conditions that would have to be present in order for life to exist.

"The main thing to remember is this," Dr. Houston continued, speaking slowly to emphasize his words. "Even if all the right conditions were present for life to exist on another planet, it still wouldn't become another earth with plants, animals, and humans. Living things didn't evolve just because all the right conditions and chemicals were present.

"They — and we — are here because we were designed and created by God, who also designed and created the earth to be the right size and distance from the sun, to have the right chemicals in the air and on the earth, and to be fine-tuned to support life in thousands of other ways.

"One more thing, Joey. Don't let anyone tell you that you aren't being smart if you believe God created humans and everything else just as Genesis teaches He did. The alternative is to believe that every living (and once living) thing on earth evolved from a one-celled organism, which would mean our ancestors included worms, fish, and reptiles. How is that more logical than God's creation?"

Discuss: What are some of the ways the earth was created that allow living things to make their homes on it?

Acids and Bases

Think about This

Roberta thought her grandmother's hydrangea bushes were the most beautiful plants she had ever seen. The flowers looked a lot like a huge snowball, except they were a sky blue color. One day her grandmother dug up a hydrangea bush and planted it on the other side of the yard. Later, Roberta noticed that the flowers were pink instead of blue, although the bush still looked healthy. What do you think happened to the bush?

The Investigative Problems

What is the difference between an acid and a base?

What is litmus paper and how do you use it?

Procedure & Observations

Taste a few of the pieces of some citrus fruits. The fruits you tasted all contained **acids**.

1. What property related to taste do they all have in common?
2. Give examples of some other foods we eat that have a sour taste.
3. Your teacher will give you some household **bases** to feel. (Don't taste these.) What was a common property of the bases related to how they feel?
4. Give some examples of other things that feel slippery.
5. Taste the water. Does it have a sour taste?
6. Feel the water. Does it have a slippery feel?
7. From what you have learned, do you think water is an acid, a base, or neutral?

Some acids and some bases are very dangerous and should not be tasted or touched. A better way to identify acids and bases is by using an **indicator**.

A common indicator is known as litmus paper. Litmus paper will not tell you if the chemical is strong or weak, but it will help you identify acids and bases. Tear your litmus paper into smaller pieces and put one piece of red and one piece of blue paper in each of the fruits or some drops of fruit juice. Do the same for the soaps and cleaners. Do the same for the water.

1. What color does litmus paper turn in fruit juice?
2. What color does litmus paper turn in the soaps and cleaners?
3. What did you observe about the red litmus paper in the water?
4. What did you observe about the blue litmus paper in the water?
5. Does this test make you more sure or less sure about your earlier decision of whether water is an acid, a base, or neutral?

The Science Stuff

Most acids and bases exist in water solutions. When acids and bases are in water, their molecules separate into positive particles and negative particles. Acids generally have positively charged hydrogen atoms. Bases form a negatively charged particle made of one hydrogen atom and one oxygen atom (called a hydroxide group). Pure water is neither an acid nor a base.

Acids and bases have different physical properties. Acids tend to have a sour taste and are found in many fruits, especially citrus fruits. Bases tend to have a bitter taste and a slippery feel. **(Remember, it is never safe to taste an unknown chemical.)**

An indicator is a chemical that changes color in acids and bases. Litmus paper is an indicator that turns red (actually pink) in acids and blue in bases. Litmus paper doesn't change color in water and other neutral materials.

Snowball plants can produce pink or blue flowers, depending on how much acid is in the soil. This is similar to how an indicator works. The acidity of the soil may be different in different locations.

A scale of numbers from 0 to 14 is used to tell whether acids and bases are strong or weak. This is known as the **pH scale**. The number seven is neutral (neither an acid nor a base). Numbers below seven are acids. Numbers above seven are bases. We will find out more about the pH scale in the next lesson.

There are many fruits and drinks that contain weak acids. There are many soaps and cleaners that contain weak bases. Strong acids and bases are very dangerous.

Making Connections

Although citrus fruits contain acids that can be eaten, some acids are dangerous to touch. Car batteries, for example, contain a strong, dangerous acid. **It is never safe to touch or taste battery acid!**

Many cleaners contain bases, but some, like oven cleaners and lye, are also dangerous. Read the labels when using strong acids and bases and heed the warnings!

The use of indicators is a much safer way to tell if an unknown chemical is an acid or a base.

Dig Deeper

Try to find out what foods and drinks contain acids.

Try to find out what common over-the-counter medications contain acids or bases.

What Did You Learn?

1. Do citrus fruits contain acids or bases?
2. Do many common cleaners contain acids or bases?
3. Are acids and bases found in water solutions or in oily solutions?
4. What happens to acids and bases when they are in water solutions?
5. Acids usually have what kind of taste?
6. Bases usually have what kind of taste and what kind of feel?
7. Chemicals that change color in acids and bases are called what?
8. What color does litmus paper turn in acids? In bases?
9. What pH numbers indicate an acid? A base? A neutral substance?
10. Are positively charged hydrogen atoms found in acids or in bases?

Basically — Is It Acid or Base?

Think about This

David and his dad are cleaning out the garage. They find a small bottle of muriatic acid, which they no longer need.

David starts to carry the bottle out when he accidentally drops it on the cement driveway. The bottle breaks and the acid spills on the driveway. David picks up a piece of the broken bottle and gets some of the acid on his hands. He immediately starts to feel a burning sensation on them.

If someone knew a few basic things about acids and bases, he or she would know what to do.

Procedure & Observations

Many household cleaners are strong chemicals that would be harmful or irritating if they got in your eyes, so wear your safety glasses for this investigation.

Your teacher will prepare the red cabbage juice the day before. (Red cabbage leaves are boiled in a pan of water. The liquid is filtered out and allowed to cool.) It will take some organization to keep up with all the different chemicals and the results, so be careful not to get things mixed up. When you have completely finished this investigation, you may try mixing the leftover chemicals to see what happens.

Pour a small amount of each of the chemicals you plan to test in small plastic cups. Label each of the cups. Put a straw in or beside each cup. Be careful not to let the straws come into contact with more than one chemical.

Now take ten more small plastic cups and put about 30 mL of the prepared red cabbage juice in each cup. Place the cups on a long sheet of paper. Write the name of each chemical you add next to the cup. Transfer the results to the data chart as you do each test. Predict whether each chemical is an acid or a base.

The Investigative Problems

How do different kinds of indicators work? Is the neutralization of an acid and a base a chemical reaction? Can an indicator help you know if acids and bases have been neutralized?

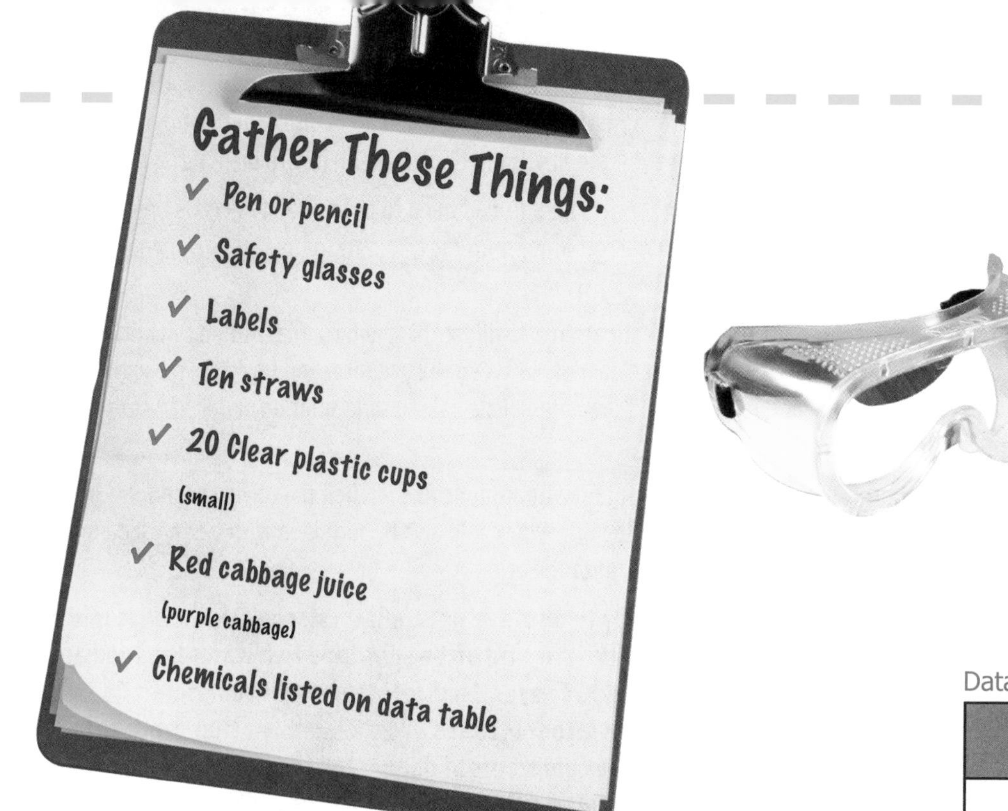

Put your first straw into the first chemical. Cover the top of the straw with your finger and move the straw over the first cup of red cabbage juice. Lift your finger to release the chemical into the juice. Observe the color and write this on the paper next to the name of the chemical you tested. Put this information in the data table. ⟶

Repeat this procedure until you have tested all ten chemicals. Be sure the data table is complete. Use the pH chart to estimate the pH number of each chemical and add this information to the data table. ⟶

Use the information in the data table to classify the chemicals you used as strong acid, weak acid, neutral, weak base, or strong base.

Find the cup with red cabbage juice and ammonia. Now use your medicine dropper to get some (pure) vinegar. Add one drop of vinegar at a time to the ammonia/cabbage juice. Count the drops. Keep adding vinegar until there is a color change. Record the number of drops you added. What color change occurred?

Find the cup with red cabbage juice and distilled water. Blow into the liquid with a straw for a few minutes. Do you observe any change in the color of the cabbage juice?

Data Table

Chemical	Predict: A or B	Color change	Estimated pH	Classify chemical
Vinegar				
Ammonia				
Lemon juice				
Distilled water				
Baking soda & H_2O				
Clear soft drink				
Liquid detergent or shampoo				
Household cleaner				
Aspirin dissolved in H_2O				
Liquid medicine for upset stomach				

pH Chart

Reds	Purples	Violet	Blue	Blue-green	Greenish-yellow
2	4	6	8	10	12

The Science Stuff

Some **indicators**, like litmus paper, can help you know if something is an acid or a base. Red cabbage juice is another indicator, but it can give you much more information than a litmus paper test. It can tell if something is a strong acid, a weak acid, a weak base, or a strong base.

In addition to red cabbage juice and litmus paper, there are many other kinds of indicators. **Phenol red** is used to test for the amount of chlorine in swimming pools by checking the acidity of the water. One of the most useful indicators is known as **pH paper**. It turns a different color for every pH number.

Red cabbage juice contains flavonoid pigments, which can change color in various acids and bases. These pigments are also found in colored fruits and flowers such as plums, grapes, blueberries, poppy flowers, and cornflowers.

The pH chart on the previous page applies to red cabbage juice color changes. It will help you know if the chemicals you tested were an acid or a base or neutral. It will also give you additional information about whether the acids and bases you tested were strong or weak.

On a pH scale, the number seven is neutral. It is neither an acid nor a base. Numbers below seven are acids, and numbers above seven are bases. The closer a number is to seven, the weaker it is. The farther away a number is from seven, the stronger it is. If a chemical turned the red cabbage juice red, it was probably a strong acid. Purples and violets were weaker acids. A greenish-yellow color probably indicated a strong base. Blues and blue-greens were weaker bases.

When you combined vinegar (an acid) with ammonia (a base), there was a **chemical change**. An acid and a base can **neutralize** each other if you put the right amounts together. This is the reaction that occurred: An acid + a base changed into a salt + water. You can tell when the solution is neutral by testing it with an **indicator**.

Another chemical change occurred when you blew air into the distilled water. Your exhaled air contains carbon dioxide. Some of the carbon dioxide reacted with the water to make a weak acid known as carbonic acid. Carbonic acid is the same acid that is found in carbonated soft drinks. You probably observed that the red cabbage juice turned a more violet or purple color as you blew into the liquid. This is because the solution changed from a neutral solution to a weak acid.

You have learned how to neutralize acids and bases. Another useful concept is **dilution**. You can dilute acids and bases by simply adding water. If you accidentally get a strong acid or a base on your skin or in your eyes, the first thing to do is dilute it with plenty of water.

By now, you should know that David and his dad need to put on safety glasses before investigating the chemicals in their garage. Since muriatic acid is a strong acid, they need to be very cautious in handling it.

The first thing to do in David's situation is to dilute the chemical that got on David's hands with lots of water. This will slow the reaction between the chemical and David's skin, so that little, if any, damage occurs. They could then put a little baking soda solution on his hands to neutralize the acid.

The broken glass should be swept up without touching it. Then the driveway and the sweeper should be washed with water. The driveway could also be sprinkled with baking soda or another weak base to help neutralize the acid.

CAUTION: You should never try to dilute or neutralize a strong acid while it is in a container. Strong acids like muriatic acid often splatter when water is added and could cause damage to your skin or eyes. Medical attention is sometimes needed, especially if a strong acid gets into someone's eyes. Remember that precautions and first aid are printed on the containers of dangerous chemicals. It's a good idea to call the local sanitation department to see how to dispose of chemicals like this that you no longer want to keep.

Dig Deeper

Do some research on acid rain and report on what you learn.

Dig up a little soil and moisten the soil with water. Use pH paper to test the soil in different places in your yard.

Identify the trees, shrubs, flowers, and grass around your home. Try to find what pH level each one needs to grow best.

Making Connections

In theory, pure water has a pH of seven, but ordinary rainwater dissolves some carbon dioxide in the air. This produces a weak acid known as carbonic acid, which causes rainwater to be slightly acidic. Slightly acidic rainwater does not cause a problem. However, sometimes nitrogen oxides or sulfur oxides are released into the air. These are classified as air pollutants because they combine with rain to form stronger acids. When they are present in large amounts, these acids are known as acid rain and can cause widespread damage to certain kinds of plants and animals on the areas where acid rain falls.

Sometimes homeowners will have the soil in their yards tested to determine its pH levels. Then they can add the right kind of fertilizers and chemicals that specific kinds of grass and other plants need to grow best. For example, adding a chemical that is a base can neutralize a soil that has too much acid.

Soap is a kind of salt. The recipe for making homemade soap involves heating fatty acids (from any kind of animal fat) with a base (generally lye).

What Did You Learn?

1. Do all soils have the same pH level?
2. What is an indicator?
3. Give some examples of indicators.
4. Can some indicators tell the difference between a strong acid and a weak acid?
5. If you accidentally spilled a strong acid or base on your skin, what is the first thing you should do?
6. What new chemical forms when there is a chemical reaction between carbon dioxide and water?
7. What new kinds of chemicals form when an acid and a base react chemically?
8. Is ordinary rainwater neutral, slightly acidic, or slightly basic?
9. Acid rain may be produced when water in the atmosphere reacts with what kinds of air pollutants?
10. What would a pH number of 7 tell you about a chemical? What would a pH of 2 indicate?

Pause and Think: Taking Care of Our Bodies

We have stressed the importance of safety in dealing with strong acids and bases. Everyone should realize that taking chemicals into your body requires good judgment, also. That includes the foods you eat, the medications you take, and illegal substances someone might offer you.

As a **Dig Deeper** project, you could do some research on a legal drug that has been abused or on some kind of illegal drug that is harmful to the human body. As you find information, keep in mind what Paul told the church in Corinth about each member's body (see 1 Cor.6:19).

Salt — An Ordinary Substance with Extraordinary Powers

Think about This

Ordinary table salt, known as sodium chloride, is only one of the thousands of salt compounds that exist on earth. Sodium chloride is one of the earth's most important compounds and one of the most abundant. It is found in the oceans, in underground salt deposits, and in all living things. At one time, the salaries of Roman soldiers were paid in bags of salt. Jesus told His disciples, "You are the salt of the earth" (Matt. 5:13). Think about why salt is both ordinary and extraordinary.

Dmitri Mendeleev organized a chart of the known elements.

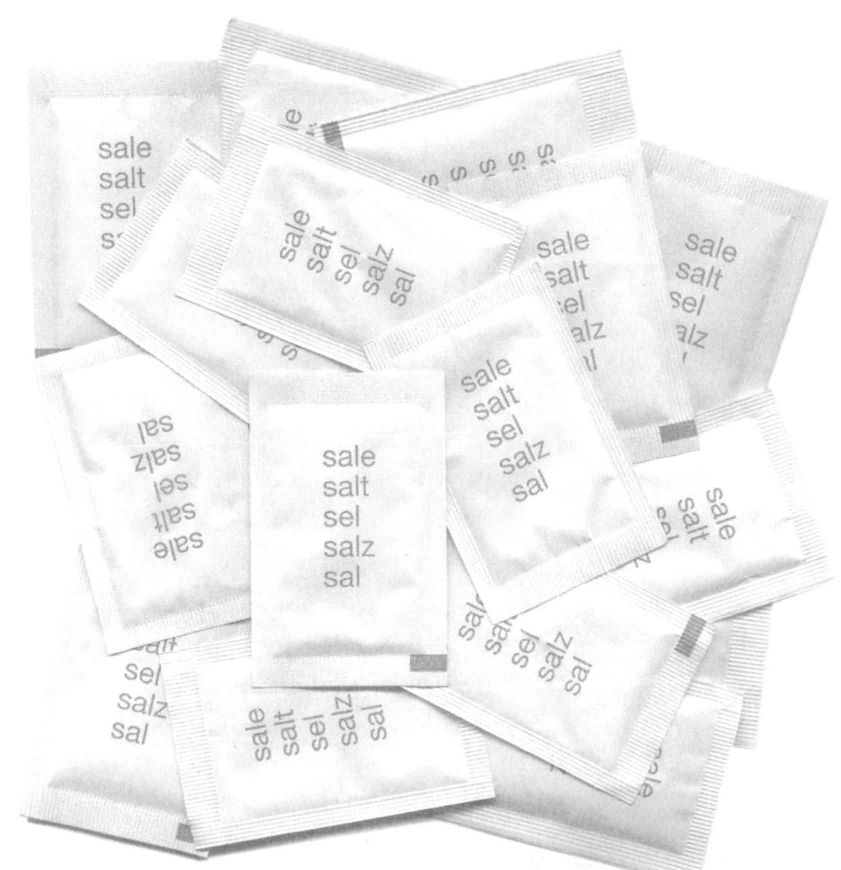

The Investigative Problems

How can you use the Periodic Table to name some common salts and write their formulas? What are some of the physical properties of table salt?

Gather These Things:

- ✓ Periodic Table of the Elements (see Appendix)
- ✓ Metal spoon
- ✓ Clear plastic cup
- ✓ Water
- ✓ Salt
- ✓ Clear glass or plastic plate
- ✓ Magnifying lens

Procedure & Observations

Examine a sample of table salt crystals with a magnifying lens. Describe the shape of the crystals (don't worry about the size).

Describe other properties of the crystals such as color and texture. Use the back of the spoon to press on some of the crystals. Are they easily crushed?

Add one tablespoon of table salt to a glass of water without stirring. For a few minutes you may be able to see some of the salt in the bottom of the glass. Now stir until the salt is completely dissolved. Do you see any evidence that there is a tablespoon of salt in the water?

Place a teaspoonful of the salt water on a glass or plastic plate. Fan the liquid until the water evaporates. Use the magnifying lens to examine the crystals that form. How do the properties of these crystals compare to the salt crystals you observed earlier?

Look at a copy of the **Periodic Table of the Elements**. This table is one of the most useful tools of chemistry. It shows all the natural elements on earth and many of the man-made ones as well. A great deal of information about every element can be obtained from the Table.

Table salt is not listed on the Periodic Table of Elements, because salt is not an element. Many of the early scientists did not realize salt was made up of two substances, chemically bonded together, because nothing in its appearance indicated this.

The elements that make up salt are the metal sodium and the nonmetal chlorine. On the Periodic Table, **metals** and **nonmetals** are two big groups of elements. Metal elements are generally shiny and solid at room temperature. They tend to be good conductors of heat and electricity. Nonmetal elements tend to have the opposite characteristics. Find the stair step dividing line for metals and nonmetals on the Periodic Table. Are there more kinds of metals or nonmetals on the Periodic Table?

The chemical name for table salt is sodium chloride. Be sure you understand how to tell that the element sodium is a metal and the element chlorine is a nonmetal by using the Periodic Table. When these elements combine chemically they form sodium chloride, a compound with properties that are different from either sodium or chlorine. The compound's name is obtained by naming the metal and changing the ending of the nonmetal to "-ide."

Use the Periodic Table to tell if the following elements are a metal or a nonmetal: rubidium, cadmium, sulfur, niobium, phosphorus, barium, and neon.

See if you can name these compounds: KCl and LiF.

The compound's formula is written by using the **chemical symbols** of sodium and chlorine instead of their names. The chemical symbol for sodium is Na and the chemical symbol for chlorine is Cl. The symbol for the metal is written first, so the **chemical formula** for sodium chloride is NaCl. There are no subscript numbers after Na or Cl, which means that there is only one atom of sodium and one atom of chlorine.

The Science Stuff

As you know, NaCl is the chemical formula for table salt or sodium chloride. The "Cl" part of the formula comes from the word chlorine. That is fairly easy to understand. But how do you get "Na" out of sodium? Well, the answer is that the "Na" comes from the Latin word for sodium, *natrium*. So, now you know.

There are many different kinds of salt other than sodium chloride. Many salts are combinations of metals (with a positive charge) and nonmetals (with a negative charge).

When sodium chloride is put into water, the bond between sodium and chlorine is broken. Sodium becomes a positively charged particle and chlorine becomes a negatively charged particle. Because water is a polar substance, it can dissolve substances that form charged particles. The following diagram shows how polar water molecules interact with the positively and negatively charged particles of salt. (See the diagrams to the right.)

Gold, silver, copper, iron, and other metals are actually elements. They are mentioned throughout the Old and New Testaments, but it was not until thousands of years later that scientists were able to tell the differences in elements and compounds. As more and more elements began to be identified, scientists tried a variety of ways of grouping them into charts and tables according to their properties. It was Dmitri Mendeleev (1834–1907) who finally organized a chart of the known elements and left blank spaces where he thought there should be elements. He correctly predicted many elements that had not yet been discovered.

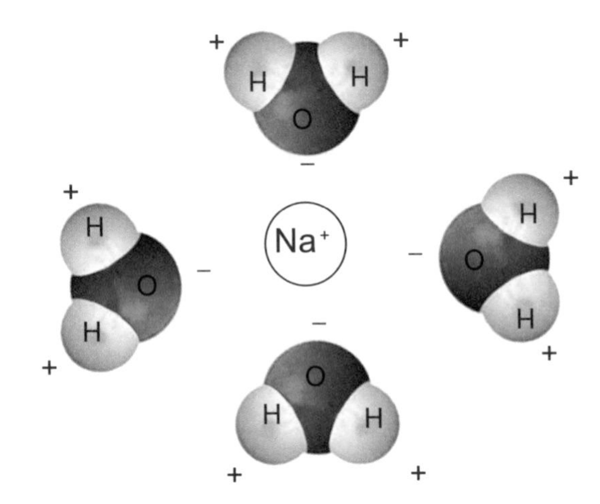

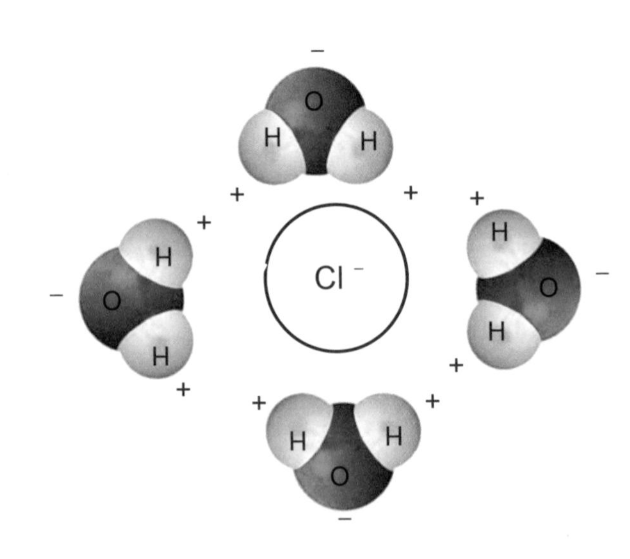

Making Connections

Most acids are made up of positively charged hydrogen atoms (H+) and certain nonmetal atoms. Most bases are made up of metal atoms and a negatively charged group of atoms containing hydrogen and oxygen (hydroxide group or OH-). Do you remember that an acid and a base can neutralize each other by forming water and a **salt**? A salt can be made up of a metal and a nonmetal. During the process of neutralization, the metal comes from the base and the nonmetal comes from the acid. Water molecules are formed as hydrogen (from the acid) and the oxygen-hydrogen pair (from the base) combine.

For example, $HCl + NaOH \rightarrow H_2O + NaCl$. Written out, this means hydrogen chloride (an acid) and sodium hydroxide (a base) change into water and sodium chloride. This kind of reaction is called neutralization, because the acid and the base neutralize each other. HCl and NaOH are dangerous acids and bases, but if they are combined in exactly the right amount, they change into ordinary water and salt.

Dig Deeper

Salts are continually being added to the ocean. Rainwater dissolves salts and other chemicals from the ground. The runoff enters streams and rivers and is eventually carried to the oceans. Presently, seawater is about 3 percent salt by weight. That translates to almost two pounds of salt in each cubic foot of seawater. If the earth is really millions and millions of years old, the oceans should be much saltier than they are now. Try to find what scientists have to say about this. Look for opinions from scientists who believe the earth is millions of years old and from scientists who believe the earth is only thousands of years old.

What Did You Learn?

1. Does the Periodic Table contain elements or compounds?
2. Where are metals found on the chart?
3. Where are nonmetals found on the chart?
4. What kinds of elements are generally found in a salt?
5. What is a chemical symbol?
6. What is a chemical formula?
7. Write the chemical formula for sodium chloride and tell what information the formula gives you.
8. List at least three physical properties of sodium chloride.
9. When an acid and a base combine in the right amounts, what do they change into?
10. Why don't you find water listed on the Periodic Table?
11. Do sodium and chlorine keep their same properties when they chemically combine to make salt?

Pause and Think: Salt of the Earth

Jesus told His small band of disciples that they were the salt of the earth. Read Matthew 5:13. Some Bible scholars believe Jesus was encouraging them to realize that even though they were few in number, they would be able to accomplish the huge tasks He gave them to do. They were like salt in that only a little salt is needed to flavor a lot of food. Do you ever feel that your life is insignificant or that one person can't accomplish much? If so, remember the effects of only a few grains of salt.

Is there a warning in the verse you read? What do you think it means?

More about the Amazing Periodic Table

Think about This

You have learned that sodium can combine with chlorine and form common salt (sodium chloride or NaCl). If you look on the Periodic Table, you will find lithium and potassium are in the same vertical column as sodium. The properties of these metals are similar to the properties of sodium, so you can predict that potassium would react chemically in the same way as sodium. Since sodium and chlorine make sodium chloride or NaCl, you would expect potassium and chlorine to make potassium chloride or KCl. This is not a salt you would want to put on your food, but it is a good example of the kind of information the Periodic Table can give you if you understand how to use it.

The *Hindenburg* disaster took place on May 6, 1937, as the German rigid airship *Hindenburg* caught fire and was destroyed within one minute while attempting to dock with its mooring mast at Lakehurst Naval Air Station in Manchester, New Jersey. Of the 97 people on board, 35 people died in addition to one fatality on the ground. The accident served to shatter public confidence, and marked the end of the giant passenger-carrying rigid airships.

The Investigative Problems

What are some ways the Periodic Table can be used to predict things about the elements?

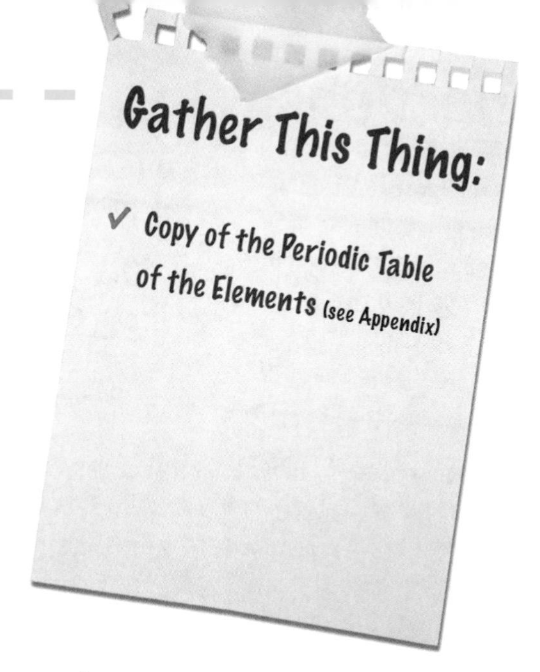

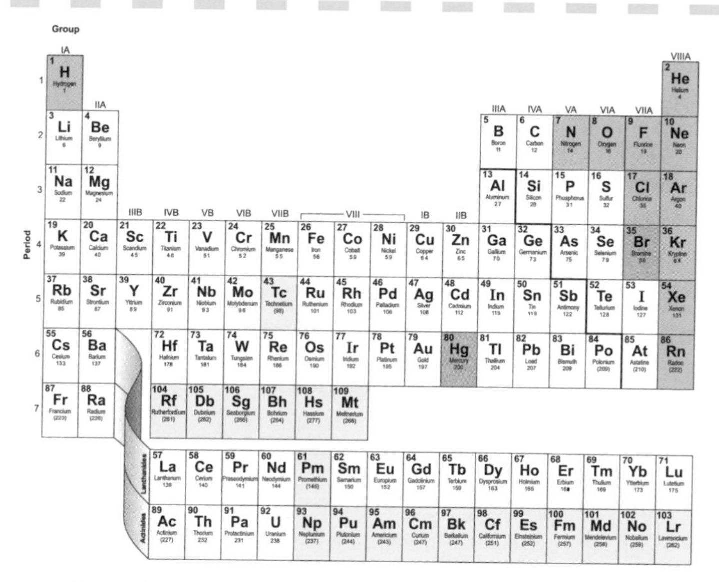

Procedure & Observations

Vertical **columns** (or families) on the chart have similar properties. Find the column that is labeled IA, which is the same column where sodium is found. Find the column that is labeled VIIA, which is the column where chlorine is found. Elements in column IA react with elements in column VIIA to form salts, with one important exception. Hydrogen reacts with elements in VIIA to form acids. Predict four salts that could result from combinations of elements from columns IA and VIIA. Name them and write their formulas using sodium chloride as your example.

Calcium is an element in column IIA and fluorine is an element in column VIIA. Calcium and fluorine can combine to form the salt calcium fluoride. It has the formula CaF_2, which means there is one atom of calcium for every two atoms of fluorine. Predict four salts that could result from combinations of elements from columns IIA and VIIA. Name them and write their formulas using calcium fluoride as your example.

Find column VIIIA. You will almost never see compounds made of these elements. These elements tend to be unreactive. They may be referred to as the inert gases in some textbooks. Name an element from column VIIIA. On the other hand, pure elements from columns IA and VIIA tend to be extremely active and dangerous, but as part of a compound, they will have completely different properties. Name an element from column IA. Name an element from Column VIIA.

Notice that each element has two numbers. The first number is the number of positively charged particles (**protons**) in one atom of that element. The second number is an average number, showing the total number of protons and neutral particles (**neutrons**) in each atom. Both protons and neutrons have mass or weight, so the more particles there are, the more the element will weigh. Look at the second number and see if you can predict which element in each pair would be heavier (assume equal sizes): iron or gold; lead or actinium; aluminum or tin; sodium or potassium; hydrogen or helium.

Look again at hydrogen and helium. They are both lightweight gases. Scientists believe these two gases are the main chemicals in most of the stars in the universe. Hydrogen is so reactive it can be explosive. Helium is the opposite and almost never reacts with anything.

During the 1930s, German engineers built two huge dirigibles or big balloons attached to a cabin where people could sit and travel from place to place. A dirigible is similar to a blimp you may see flying over football games on TV. Unfortunately, they chose hydrogen gas to fill the balloons. One passenger balloon named the *Hindenburg* was involved in a terrible accident where a spark caused the hydrogen gas to react with oxygen in the air and explode. Use the Periodic Table and predict what might have happened if the *Hindenburg* had been filled with helium gas instead.

Find the dividing line between metals and nonmetals. Recall the general properties of metals and nonmetals. Predict four elements that would conduct electricity and four elements that would not conduct electricity. Except for aluminum, the elements that are touching the dividing line can be tricky. They tend to have nonmetallic properties, but can conduct electricity as semi-conductors. Another exception is hydrogen. It is listed on the metals side, but it has general properties of a nonmetal.

The Science Stuff

The Periodic Table contains all the natural elements in the universe plus some of the man-made ones. Each element has its own name, chemical symbol, number of protons, and average weight of each atom.

A dividing line separates metals and nonmetals. Metals are generally shiny, are good conductors of heat and electricity, and can be easily hammered without breaking. Nonmetals usually are not shiny, do not conduct heat and electricity well, and tend to break or crumble when hammered. Metals are found on the left side and nonmetals are found on the right side of the dividing line. Hydrogen, a nonmetal, is found on the left side. Except for aluminum, most of the elements touching the dividing line tend to have both metallic and nonmetallic properties.

Acids often contain a nonmetal (with a negative charge) and bases often contain a metal (with a positive charge). When an acid and a base react, the metal from the base and the nonmetal from the acid combine to make a salt.

You should recall that acids also have a positively charged hydrogen atom. Bases have a negatively charged group composed on one oxygen atom and one hydrogen atom. These elements combine to form water.

Thus, an Acid + a Base ⇢ Water + a Salt. Remember, the arrow can be read "changes into."

Groups of elements in vertical columns tend to have similar chemical properties, and often have similar physical properties. For example, the metal elements in column IA tend to be very reactive, often explosive. The nonmetal elements in column VIIA also tend to be very reactive and dangerous. Elements in column VIIIA almost never react with anything.

Formulas for molecules can be written from the symbols of the elements in the compound. The number of atoms in a molecule is written as a subscript after the symbol. If there is no number after a symbol, it means there is one atom.

There is a great deal of other information about the elements that the Periodic Table can give you. As your knowledge of chemistry expands, you will be even more amazed at how useful this tool can be.

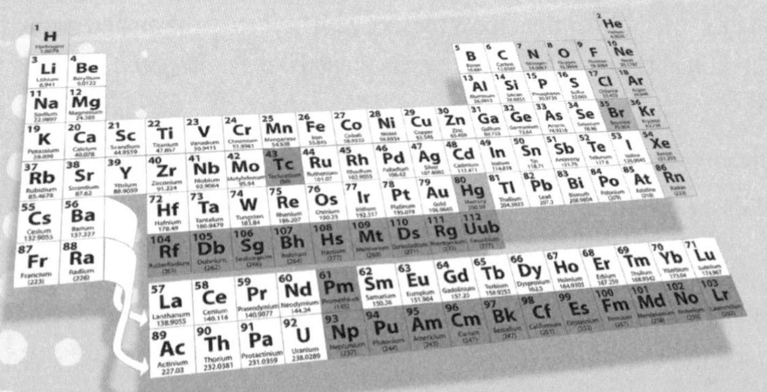

Making Connections

We know today that water is a compound made up of hydrogen and oxygen; that air is a mixture of gases; that the earth is a mixture of many kinds of substances; and that fire is the point where a substance is rapidly combining with oxygen and giving off heat and light energy. It took many scientists many years to finally understand water, air, earth, and fire and to discard the old theory of four elements.

We should be careful not to think that earlier scientists weren't very smart simply because they didn't know the things we know today. Try to place yourself in the 1500s or 1600s as a part-time scientist studying water. What would you do to see if water was made up of simpler substances?

Dig Deeper

By the time Jesus was born, smelters of metals had found seven purified metal elements — gold, silver, copper, iron, lead, tin, and mercury. They didn't understand how these metals related to the philosopher's four basic substances, but they knew how to treat the ores so that the metals could be extracted. They also knew about charcoal (carbon) and brimstone (sulfur). Do some research about how the ancient smelters recovered metals and about how the discovery of metals affected history.

Learning to separate iron from iron ore was such an important discovery in history that the first uses of iron began what is called the "Iron Age." What were some of the history-making ways in which iron was first used by early civilizations?

Try to find an audio or video account of the *Hindenburg* crash on the Internet or from some other source. Why do you think this was such a huge news story at this time?

What Did You Learn?

1. The Periodic Table consists of a series of blocks containing symbols and numbers organized into columns and rows. What do these blocks contain information about — elements or compounds?
2. Where do you find similar groups of elements — in vertical columns or in horizontal rows?
3. Except for hydrogen, do you find metals or nonmetals on the left side of the Table?
4. Where are most nonmetals found?
5. Predict: Which element in each pair would be more dangerous or more reactive — potassium or calcium; sulfur or chlorine; krypton or selenium?
6. Predict: Would astatine have properties more like radon or like iodine?
7. Which scientist organized the known elements into a Periodic Table and left blanks where he predicted undiscovered elements would go?
8. Predict: Would a block of gold weigh more or less than an equal size block of uranium?
9. Predict: Which element would be the best conductor of electricity — palladium or phosphorus?
10. Predict: Which element would be a shiny solid — rhenium or krypton?
11. What do an acid and base change into when they react chemically?

Pause and Think: Metals in the Bible

Use a Bible concordance and try to find how many metals are mentioned in the Bible. Which of these metals are mentioned in the Book of Genesis? Name each metal and give at least one Bible reference for each metal you named.

Dumping a Bad Theory

Once most scientists accept a scientific theory, it is difficult to do away with it. For about two thousand years, most philosophers and scientists thought there were four elementary substances from which everything else was made. These were fire, air, water, and earth. It was thought that they were in their simplest form and could not be broken down into anything else.

A version of this theory was accepted by most scientists/philosophers until sometime around the 1600s. During that time, a few European scientists, like Robert Boyle, began to doubt that fire, air, water, and earth were the only four basic elements. But the theory did not die easily or quickly.

Robert Boyle (1627–1691) was born into a wealthy family. The family wealth gave him the advantage of having time to study science. He was well-educated and even traveled to France with a tutor as a teenager where he studied with the elderly Galileo Galilei in 1641.

Boyle's name is found in almost every chemistry book, because he was the first to mention that there is an opposite relationship between the volume of a gas and the pressure of the gas. This famous principle is known as Boyle's Law and is usually expressed mathematically.

Robert Boyle

In 1661, Boyle published an influential book called *The Sceptical Chymist*. (This was the correct spelling in 1661.) In this book, he laid the foundations for the atomic theory of matter. He argued against limiting chemical elements only to fire, air, water, and earth. He also argued that scientists should back up their theories experimentally.

Although he loved studying chemistry, Boyle also loved to study the Bible. He believed that science and Scripture were harmonious. *A Discourse of Things above Reason* was another book he wrote and published in 1681. He believed there were limits to reasoning about what God can or cannot do, that God's attributes can be seen by studying nature, and that His wisdom can be seen in creation.

Eventually, the four basic elements theory was updated to explain new research. Burning came to be explained in terms of a particular kind of "earth" that had a mysterious element, phlogiston, in it. Burning objects were said to give off phlogiston. The phlogiston theory explained many things about burning and was perfectly logical for many more years.

Read this complete narrative at: www.masterbooks.org to see what Henry Cavendish, Antoine Lavoisier, Joseph Priestley, and John Dalton did to dump a bad theory.

Electricity and Salt Water

Think about This

Water is such a stable compound, it was considered one of the four basic elements for hundreds and hundreds of years. We know today that water is not an element at all, because it can be broken apart. One way of separating water is by using electricity. We will add salt to the solution, because pure water will not conduct electricity.

The Investigative Problems

How is liquid water changed by an electric current passing through it?

Procedure & Observations

1) Put two or three teaspoonfuls of Epsom salt in a glass beaker (or a widemouthed container). Add 250 milliliters of distilled water to the salt and stir until the salt is dissolved.

2) Fill two test tubes with water. Put your thumb over the end of one of the test tubes and put it into the beaker upside-down. The water will remain in the test tube. Repeat for the second test tube.

3) Take the alligator clips and strip about five cm of insulation from the plain ends of the wire. Bend the ends of both wires, so that the bare wire can go into the test tubes.

4) Place the bent ends of the wires in the salt water and insert them into the open ends of the test tubes. Connect the alligator clamps to the positive and negative terminals of the battery.

5) Make careful observations of what you see. Record your observations. Note which wire has the most bubbles coming off of it.

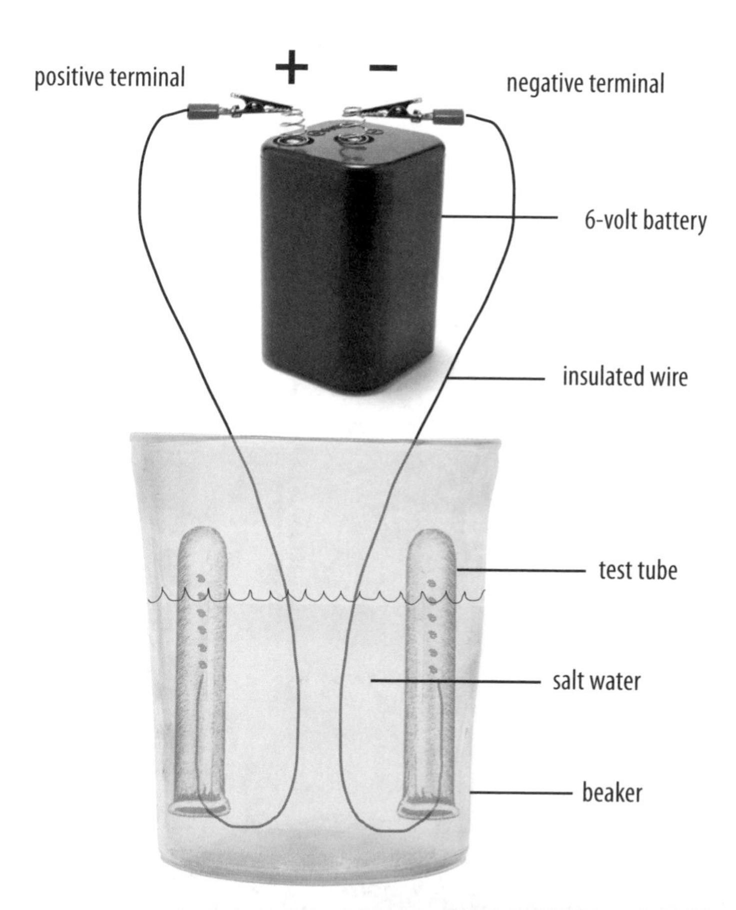

The Science Stuff

A **molecule** of water is composed of two hydrogen atoms and one oxygen atom. The atoms are bound together by strong chemical bonds. In a process called **electrolysis**, water can be separated into its two elements. The first step is the formation of positive and negative particles. As the reaction continues, these particles are finally changed into molecules of hydrogen gas (H_2) and oxygen gas (O_2).

Pure, distilled water is not a good **conductor of electricity**. In order to send an electric current through water, a chemical like Epsom salts must be added.

When the wires are connected to the battery and placed in the salt solution, gas bubbles immediately begin appearing on the bare ends of the wires. Bubbles of hydrogen gas form at the wire connected to the negative terminal. Bubbles of oxygen gas form at the wire connected to the positive terminal.

During electrolysis, water changes into the elements hydrogen and oxygen. These elements are gases. As they form, they begin pushing the water out of the test tubes. In only a few minutes, you can tell that bubbles form much faster in one of the test tubes than they do in the other one. This is because there are twice as many hydrogen bubbles produced from water as there are oxygen bubbles. The formula for water, H_2O, tells you that there are two hydrogen atoms and one oxygen atom in a molecule of water.

Dig Deeper

Electroplating is a process similar to electrolysis, and can be done with a 6-volt battery. You can electroplate a metal key with a layer of copper by connecting a piece of copper metal to one wire and a metal key to the other wire. The liquid is a concentrated solution of copper sulfate with a few drops of sulfuric acid added. It generally takes about 15 minutes to produce a copper-coated key. If you want to try this, be sure to wear safety glasses, use proper precautions, and have an adult present to help handle the sulfuric acid. **Never try this without adult supervision!**

Silver- and gold-plated jewelry can also be produced by the process of electroplating. The process of electrolysis can also make gold-plated and silver-plated jewelry. Do some research on electroplating jewelry. What are the symbols for electroplated gold and silver? How durable is electroplated jewelry? Share what you learned with others.

Making Connections

An electric current flows easily through metals, such as copper wire. It does not flow easily through things such as rubber, cork, or dry wood. Although it does not flow through pure water, it will flow through salt water.

The old fire, air, water, and earth theory would have been easily discarded years earlier if scientists had understood the process of electrolysis of water. This would have shown that water could be broken down into two other substances. You should remember that according to the old theory, water was thought to be an elemental substance that could not be broken apart into simpler substances.

What Did You Learn?

1. What is electrolysis?
2. When water is separated by electrolysis, what gas forms at the positive electrode?
3. When water is separated by electrolysis, what gas forms at the negative electrode?
4. Are there more hydrogen or oxygen bubbles formed?
5. What was the purpose of adding the salt?
6. Will pure water conduct an electric current?
7. Suppose someone in the 1500s had invented some method to separate water into oxygen and hydrogen gases. This would have been strong evidence against what popular theory?

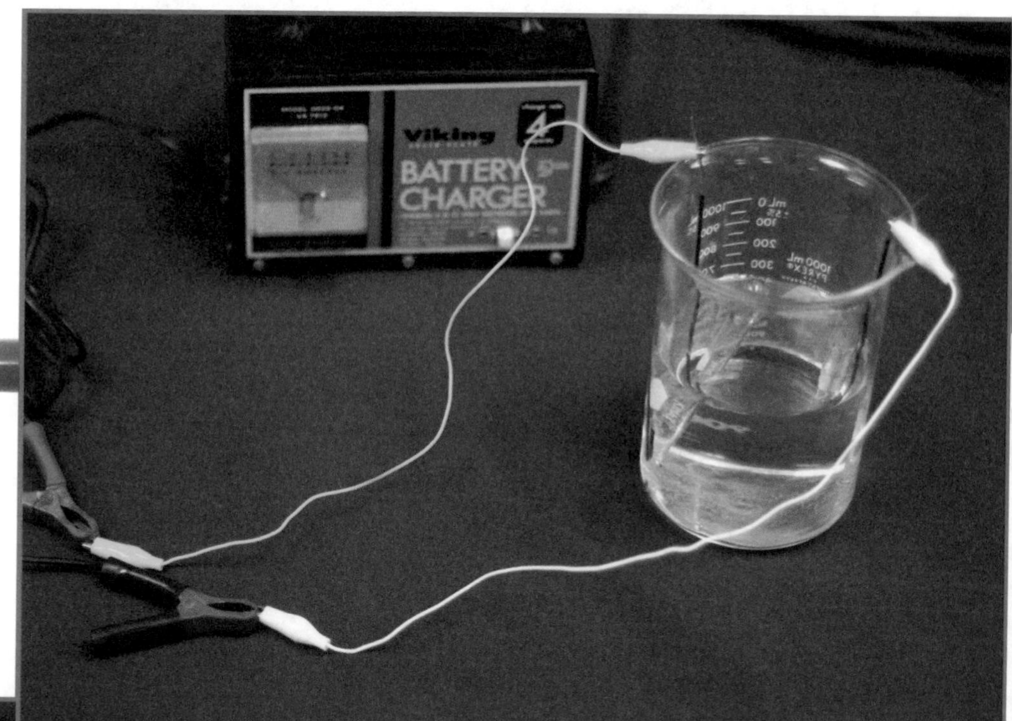

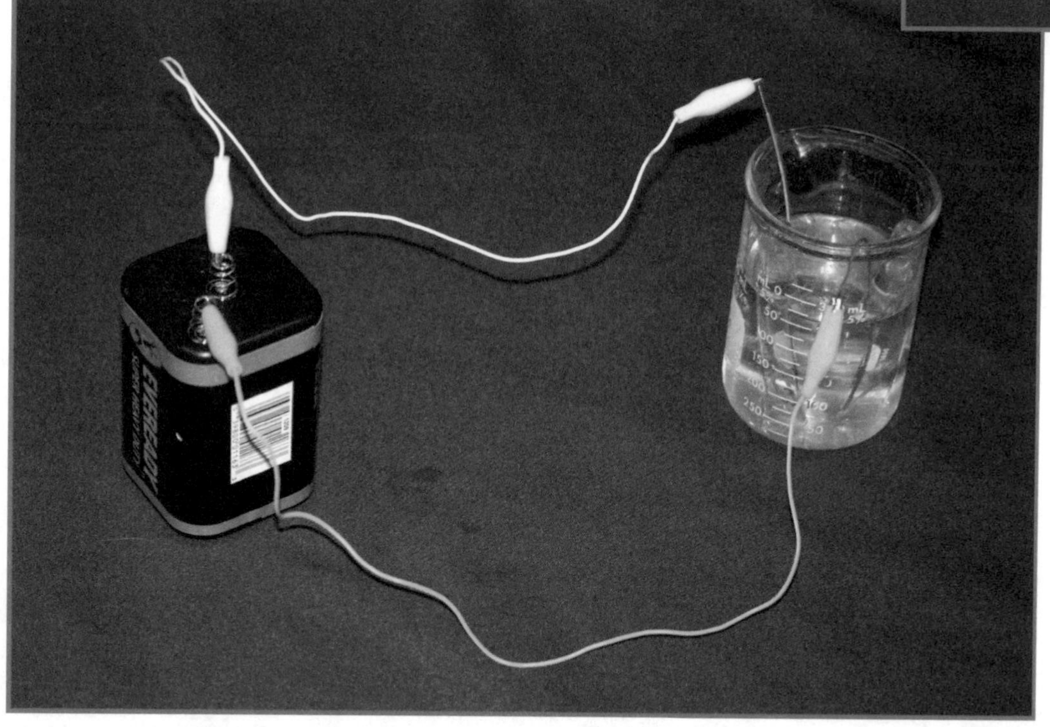

Changes — Are They Chemical or Physical?

Think about This

Have you ever seen large inflatable balloons made into cartoon figures? They may look huge and solid, but they are mostly full of air. The Styrofoam cup your teacher is holding is somewhat similar to the inflatable figures. Observe the demonstration and see if you can figure out in what ways they are similar and in what ways they are different.

Observe as your teacher puts a Styrofoam cup in a small container of acetone. Notice what happens to the Styrofoam. Do you see any bubbles coming out of the cup? Predict how many cups will go into the container of acetone as your teacher adds more cups.

Now observe as your teacher dips a piece of wood that has been painted with fingernail polish into a container of acetone.

You have just observed a chemical change and a physical change. As you do the activities in this lesson, you will gain a better understanding of the differences in physical changes and chemical changes.

CAUTION: Acetone Is Flammable And Must Be Kept Away From Flames. Vapors Are Toxic, And You Need Safety Glasses When Working With Acetone. If acetone is spilled on wooden floors or painted surfaces, it may damage the surfaces.

The Investigative Problems

What kind of change occurs when sugar is heated for several minutes? Does this result in a physical change or a chemical change?

Procedure & Observations

Activity 1:

A teacher will do this activity while students observe. While waiting their turn, other students may conduct some experiments at their desks.

Place two spoonfuls of sugar in a pan that is lined with heavy-duty aluminum foil and place on a heating plate. Everyone will make careful observations while wearing safety glasses. Notice the smells. Continue heating until the sugar changes into a substance that is black and puffy. Use tongs to move the pan to a place where it can cool. When the black, puffy material is cool, feel it to determine its texture and strength.

Record what you see, smell, and feel. Be as detailed as you can.

Can you make steam (actually water vapor) and the black substance (actually carbon) go back together and remake sugar? Try to combine a little water and carbon in another container and see what happens.

Activity 2:

Observe a mixture of sugar and sand with a hand lens and record your observations. Can you identify the grains of sugar? How are they different from grains of sand? Separate a few of each kind of grains into two groups. Can you crush either of them? Add a few drops of water with the spoon to both groups. Will either of them dissolve in water?

Activity 3:

Add one teaspoon of sugar to ¼ cup of water; stir and add another teaspoon; repeat to see if it will dissolve five or more teaspoons. The sugar and water mixture takes on a syrupy state. This mixture can reform sugar crystals, but it is a slow process and might take several days under special conditions. Compare the properties of the sugar and water mixture to the properties of pure water.

The Science Stuff

Mixing sugar and salt together, crushing grains of sugar, dissolving salt in water, melting, evaporation, boiling, condensation, and freezing are all examples of **physical changes**. Tearing up a piece of paper is also a physical change, because small pieces of paper have the same chemical makeup as a big piece of paper.

Chemicals do not lose their primary physical characteristics even when they are mixed with other chemicals. One or more of these physical properties can usually be the basis for separating a **mixture**.

You know that ice melts when it is heated, but both ice and liquid water are just two states of the same chemical. The **chemical formula** for water, H_20, is still the same.

Sugar molecules are composed of atoms of carbon, hydrogen, and oxygen. As sugar is heated, its molecules begin to break apart. The smoke you saw contained some water (hydrogen and oxygen) in the form of a white steam and some vaporized carbon (a black smoke). The black material that was left in the pan was carbon deposit, the same substance that is left when many things burn. Some carbon dioxide and carbon monoxide molecules also formed, but they mixed with the air and you couldn't see them.

Sometimes molecules break apart during a chemical reaction, as the sugar molecules did. Another kind of chemical reaction occurs when two substances join to make a new chemical, as when iron combines with oxygen and changes into iron rust. Other times, one or more atoms are replaced with different atoms.

When Styrofoam is placed in acetone, some of the weak bonds (known as cross-linked bonds) that connect the long polymer chains together are broken. The stronger bonds that make up the polymer chains are not broken. The bubbles you saw as the cups were put into acetone formed as trapped air molecules were released. Most of the changes you observed were actually physical changes.

There are chemical reactions that can be reversed under the right conditions, but it is not always easy. Trying to make or remake sugar from carbon (or carbon dioxide) and water would be extremely complicated. When iron and oxygen combine chemically, they change into iron rust. Iron rust can be made to change back into iron and oxygen, but it requires a lot of energy.

In a chemical reaction, the original chemical or chemicals have certain properties. When new chemical(s) form, they have a different set of properties from the original chemicals.

In both physical and chemical changes, the new substances weigh the same as the original substances. For example, if you weighed a teaspoon of sugar and ¼ cup of water, and then mixed them together, the sugar water solution would weigh exactly the same as the two substances you started with.

Imagine that you could weigh the sugar you started with, plus a little oxygen from the air that joined in. Then imagine that you could collect and weigh the steam, black smoke, invisible gases, and charcoal that formed. The substances you started with would weigh exactly the same as the substances that were formed after the sugar was heated. The same atoms are present before and after a chemical change, but they are rearranged to make new chemicals.

Dig Deeper

What is the difference between glucose and sucrose, two forms of sugar? Find the chemical formulas for both compounds and write them in your Student Answer Book. What products are formed in your body when oxygen combines with a sugar that you have eaten? Is energy given off when oxygen combines with a sugar?

Make a list of five chemical changes and another list of ten physical changes.

Making Connections

Did you know you would become extremely wealthy and famous if you could find a simple, inexpensive way to make carbon and water combine to form sugar? Yet every green plant has the ability to put carbon, hydrogen, and oxygen atoms together to make a simple sugar and give off oxygen gas as a by-product. They do this by combining carbon dioxide and water in the process of photosynthesis. This is a very complicated process involving many steps. Scientists have learned much about these processes, but not enough to build factories and duplicate them in an economical way. All living things contain carbon. All foods that come from plants and animals contain carbon. All fuels that were made from plants or animals contain carbon. The elements of oxygen and hydrogen are also present in all living things.

What Did You Learn?

1. Give several examples of physical changes.
2. Give several examples of chemical changes.
3. Read the definitions of both physical and chemical changes. Tell the differences in your own words.
4. Name the kinds of atoms that make up a sugar molecule.
5. What is the black substance that is left after sugar decomposes?
6. What was the name of the white smoke that formed from hydrogen and oxygen atoms released from the sugar?
7. Was the decomposition of sugar a chemical or a physical change?
8. Is dissolving sugar in water a chemical or a physical change?
9. Give two ways in which sugar and sand have different physical properties.

Find examples of chemical change on this page.

Find examples of physical change on this page.

Mowing the yard is an example of physical change.

Clues of a Chemical Reaction

Think about This

Molly thought it was confusing trying to tell the difference between physical and chemical changes. She knew chemical reactions were changes that produced new chemicals. She found it helpful to look for the clues of a chemical reaction: temperature changes, bubbles, color changes, and chemicals that won't dissolve. Keep these clues in mind as you do these activities.

Combustion is an example of a chemical change.

Procedure & Observations

Part I:

Place about ½ cup of water in a clear glass container. The water should be room temperature. Feel the container. Add a piece of an effervescent tablet to the water. After about 15 seconds, feel the container again. Is it warmer or cooler than it was before?

Write your observations of what happened when you added the effervescent tablet to the water.

What are two clues that would make you think a chemical reaction had taken place?

The Investigative Problems

What are some examples of chemical reactions? What are some clues that a reaction has occurred? What clues do you observe?

Part II:

Insert a straw into the cup lid. Seal the space around the straw with putty or soft modeling clay. Fill the cup half full of water. Prepare a second cup in the same way, but add a few drops of Phenol Red to the water. Put the lid on the second cup. Push the two straws together and connect them with a piece of tape, so that there won't be a leak. Put ½ of an effervescent tablet into the first cup. Quickly put the lid and straw over the opening.

As much as you can, observe evidence of bubble formation and color changes. Note any sounds you hear as well. After a few minutes, remove the lid to the cup with the Phenol Red. What color is the solution now?

Can you tell that changes occurred in both cups? Record your observations.

What clue would make you think a chemical reaction had taken place in the cup with the Phenol Red?

Part III:

Your teacher will give you two liquids. One of these was produced by soaking three tea bags in a steel wool pad in a cup of vinegar overnight. The other one is a strong solution of tea made by steeping a cup of hot water. After recording your observations of the two liquids, mix a little of each in a clear plastic cup. What clue would make you think a chemical reaction had taken place? (Caution: Be careful when handling this. It is like permanent ink that doesn't wash out of clothing!)

Part IV:

Add a spoonful of liquid dishwasher detergent to a half cup of water. Add a spoonful of Epsom salts to another half cup of water. Combine the two solutions in a clear plastic cup. Observe how the two solutions look before they are combined. Then observe how the solution looks after they are combined.

The Science Stuff

Both physical and chemical changes occur all around us. In **physical changes**, the substance keeps its primary properties. The arrangement of the substance's atoms and molecules remain the same. In **chemical changes**, there is a rearrangement of the **atoms** and **molecules**. These new substances will have different properties than the original chemicals.

There are four clues that are often present when a chemical change (reaction) takes place. These clues are a temperature change, formation of bubbles, a color change, and the formation of a chemical that won't dissolve in the solution. Changes in chemical indicators may also be a clue of a chemical change.

When you added an effervescent tablet to water, the temperature changed and the liquid became colder. At the same time, you saw bubbles rising to the surface, indicating that a gas had formed during the reaction.

When effervescent tablets and water react, carbon dioxide gas is formed. You should remember that carbon dioxide and water react to form carbonic acid, a weak acid. When carbon dioxide is transferred to a solution of water and Phenol Red, carbonic acid begins to form. This causes the Phenol Red indicator to change colors. The color change in this case showed that the solution changed from a neutral solution to a weak acidic solution.

Soaking a steel wool pad in vinegar frees up particles of iron and keeps the iron from combining with oxygen to make iron rust. When the tannic acid in the tea combines with iron, it forms a compound called ferrotannate, similar to a chemical that was used for hundreds of years as ink for quill pins. The color change is a clue that a chemical reaction took place. **Be careful not to get this chemical on your clothes!** (Optional: Use a cotton swab to write something on a piece of paper with the steel wool solution. After its dries, put another cotton swab in the tea and rub all over the paper. The writing will appear. Let this dry and see if you can wash the writing off with a little water.)

When a solution of Epsom salts is combined with liquid dishwasher detergent, a new chemical forms that doesn't dissolve in the solution. The formation of an insoluble chemical is another clue that a chemical reaction occurred.

Dig Deeper

The chemical that was used for many years as an ink for writing on paper was ferrogallotannate. It was produced by grinding up oak galls with iron in a liquid solution. The oak galls contained gallic and tannic acids. The ink darkened as it further reacted with oxygen in the air. Various writing instruments have been used in American history. See if you can find more about what was used in the 1700s to write letters, to make documents, and to do homework.

Hundreds of years ago, the field of chemistry was advanced by an unlikely group known as alchemists. They performed many experiments trying to turn iron and other metals into gold. Even though the alchemists were wrong about many of their ideas, they still contributed to the field of science. See if you can find some of the ways they helped to advance science.

Put an unbroken, uncooked egg in a pint of vinegar. Put another unbroken, uncooked egg in a pint of water. Observe for several minutes. Leave the eggs in the liquids for a day or two until the shell of one of the eggs is removed. Take the eggs out of the containers and observe them carefully while wearing safety glasses. Write your observations. The eggshells are made of calcium carbonate. Find the name of the acid in vinegar. On the basis of what you learned in this lesson, in which container was there a chemical reaction? Tell why you think this. Try to determine what changes may have taken place. **CAUTION: Raw eggs that are left outside of the refrigerator can be a source of bacterial growth. Wash your hands with anti-bacterial soap any time you contact the eggs and dispose of the eggs properly when you finish your experiment.**

Making Connections

There are chemical changes all around you. Burning a candle, the rusting of iron, baking a cake, and combining a metal with an acid are all examples of chemical changes. Although you can't see this, the process of digestion in your body involves a series of chemical changes. Other chemical changes also occur as your body combines food and oxygen to release energy. Green plants combine carbon dioxide and water in a complex series of chemical changes to make glucose and other foods. The processes of making nylon, synthetic rubber, Styrofoam, and plastics also involve chemical changes.

What Did You Learn?

1. State which of the following changes are chemical changes: boiling water, freezing water, adding vinegar to baking soda, dissolving salt in water, combining an acid and a base.

2. A forensic scientist wants to know if a certain chemical is present. When a few drops are added to a solution, an insoluble substance forms that is a bright yellow. Is it likely that a chemical change occurred?

3. What is the difference between a physical change and a chemical change?

4. A piece of zinc is added to a test tube containing hydrochloric acid. Bubbles start to rise in the tube and the tube begins to feel warm. Is it likely that a chemical change occurred?

5. Suppose you have an unknown gas and you allow a small amount of the gas to bubble through a solution of limewater. What would happen if the gas was carbon dioxide?

6. What are four clues that a chemical reaction has taken place?

A Heavy Gas

Think about This

Kayla lit the candle, blew out the match, and pitched the burnt match in the trash. A few minutes later, she told her dad that she thought she smelled smoke. At first they thought it was just the smell of the burning candle, but when they went in the kitchen to check on everything, they discovered a blazing fire in the trash can. Kayla's dad quickly grabbed the fire extinguisher and sprayed over the fire. In seconds, the blaze was out. Kayla thought it looked like an invisible blanket had been dropped over the trash.

A fire extinguisher is a handheld device used to extinguish or control small fires.

"Your match looked like it was out, but there was still a glowing ember inside that caught the paper on fire. Our fire extinguisher produced a heavy gas that just settled over the blaze and cut off all the oxygen. Fires have to have a fuel, high temperature, and oxygen to keep burning. When the oxygen is cut off, there's no more fire," her dad explained.

What do you think the heavy gas was?

The Investigative Problems

How can we tell that a gas is produced by a reaction between vinegar and baking soda?

Procedure & Observations

Use a funnel to guide about two tablespoons of baking soda into a balloon. Add about ¾ cup of vinegar to the empty two-liter plastic bottle. (You may add red cabbage juice or another indicator to the vinegar if you want to see when the vinegar is neutralized.) Attach the balloon to the opening of the bottle, but don't let the baking soda fall into the vinegar yet. When the balloon is stretched securely over the bottle opening, hold the balloon up and allow the baking soda to fall into the vinegar. Carefully observe what happens and write this in your notebook. Include measurements of the balloon and the time the reaction lasted.

Place a small candle (which is sitting in a small heavy holder) in a deep bowl. Add a few inches of water around the holder, depending on the height of the candle. Leave two or more inches of the candle above the water. An adult will light the candle and let it burn for a few seconds. Then lower the drinking glass over the candle and into the water. Don't let the glass get too close to the candle. Observe what happens.

Pour out the water and replace it with vinegar. Put the candle and holder back in the bowl. Have an adult light the candle again and let it burn for a few seconds. Add about a tablespoon of baking soda to the vinegar. Carefully observe what happens to the flame. (If the liquid puts out the flame, try again with a smaller amount of vinegar.) Record all observations in your notebook.

When you finish, list as many properties of the gas that formed when vinegar and baking soda reacted as you can.

The Science Stuff

A **chemical change** occurs when baking soda and vinegar are combined. The formation of bubbles is a clue that this was a chemical change. You can assume that a gas formed, because you could see bubbles rising from the bottle and the balloon expanded. If you put an indicator in the vinegar, you would have observed that the vinegar was an acid; when the baking soda was added, the solution became almost neutral.

Vinegar is made up of the elements hydrogen, carbon, and oxygen. Baking soda is made up of the elements sodium, hydrogen, carbon, and oxygen. The gas that bubbled out of the liquid was carbon dioxide (CO_2).

The reaction proceeded rapidly, adding more carbon dioxide to the air in the bottle and the balloon. As soon as the two chemicals finished reacting, the bubbles stopped forming.

You should have been able to extinguish the candle flame by putting the burning candle in a bowl surrounded by vinegar and adding baking soda to the vinegar. The carbon dioxide that formed from the reaction filled the bowl and acted like a blanket to cut off the oxygen supply in the bowl. Without a source of oxygen, the candle went out.

The candle will also go out when a drinking glass is placed over the candle and into water. As soon as the oxygen inside the glass is used up, the candle will go out. In order for any substance to burn, it must have oxygen. As soon as the oxygen is cut off or used up, it will stop burning.

Carbon dioxide is a colorless, odorless gas that doesn't burn. It is also a heavy gas that tends to sink down when released. These properties make carbon dioxide a good choice for a gas to be produced in a fire extinguisher. Lighter gases like hydrogen and helium tend to rise up quickly when released.

Making Connections

You learned that H_2O is a **polar molecule**, because it is slightly positive on one end and slightly negative on the other end. This is partly because the two hydrogen atoms are not located opposite each other. In the case of CO_2, the charges are balanced. This is because the two oxygen atoms are on opposite sides of the carbon atom. Therefore, CO_2 is not a polar compound.

Green plants take in and use carbon dioxide as they make food.

Carbon dioxide can be made into a solid if its temperature is lowered to almost -80° C. The solid state is known as "dry ice" because it doesn't melt. It just changes directly from a solid to a gas. If you caught some fish and needed to keep them cold for several hours while traveling home, you could try this trick. Wrap the fish in paper and place them in a covered cardboard box with some dry ice. The fish will be cold and the box and paper will be dry when you get home — no mess. However, you would need to take some precautions with the dry ice. It is extremely cold and could freeze your skin if you picked it up. It could also push the oxygen out of a small room and make it hard to breathe.

Dig Deeper

Green plants use carbon dioxide and water to make food (in the form of sugar). Animals exhale carbon dioxide and water vapor after they have eaten food. The burning of coal, oil, gas, and wood also gives off carbon dioxide and water vapor. Some scientists believe carbon dioxide is entering the atmosphere faster than it is being taken out. Do you know why some people are concerned about too much carbon dioxide accumulating in the air? They fear that the extra carbon dioxide in the air may be causing the earth to become warmer, in a process known as global warming. Try to find at least two different scientific opinions about global warming. Why do different scientists have different opinions about this when they are all looking at the same evidence?

Carbon monoxide is another compound that contains carbon and oxygen. Its formula is CO. It is very harmful to breathe and can even cause death if it is breathed too long. It may be produced when fuels burn incompletely. For example, it is a by-product of burning gasoline in car engines. It can also be produced when gas-burning heaters are not regulated properly. (Yellow flames with black soot deposits may indicate the heater isn't getting enough oxygen.) What are the symptoms of carbon monoxide poisoning? What are some safety rules to prevent carbon monoxide poisoning?

What Did You Learn?

1. What two things did you observe that would make you think a gas was being produced in the activity you did?
2. What was the name of the gas that was produced when vinegar and baking soda reacted?
3. In order for any substance to burn, what gas must be present?
4. List three properties of carbon dioxide.
5. What is "dry ice"?
6. Does "dry ice" melt when it gets warmer like ordinary ice does?
7. Explain why a burning candle will go out when carbon dioxide is produced around it.
8. Explain why a burning candle will go out when a glass is placed over it.
9. A green plant takes in and uses what gas as it makes food?
10. Is carbon dioxide a polar compound?

Do not do experiments without parental supervision.

Large or Small? Hot or Cold?

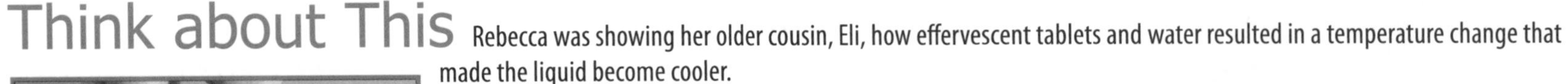

Think about This

Rebecca was showing her older cousin, Eli, how effervescent tablets and water resulted in a temperature change that made the liquid become cooler.

"This is a very fast reaction, isn't it?" Rebecca said.

"Bet I can make it go faster in two ways," Eli responded.

What do you think? Does Eli know what he's talking about?

The Investigative Problems

Does particle size affect the rate of reaction time?

Does temperature affect the rate of reaction time?

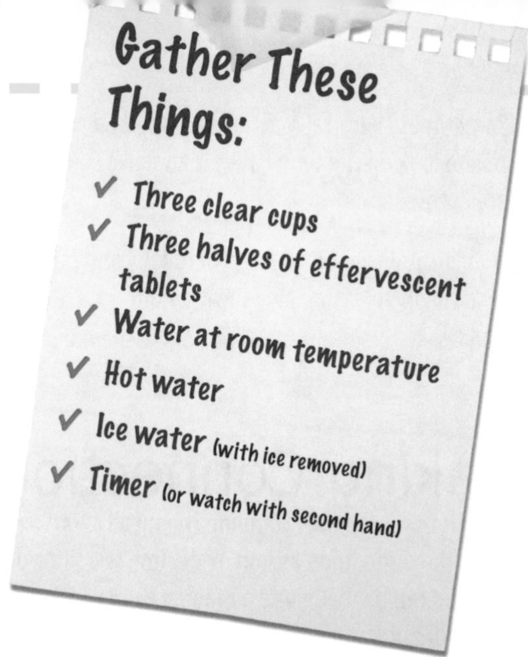

Procedure & Observations

Part I: Place labels under the three cups to identify fine powder, small pieces, and whole piece. Grind one of the half tablets into a fine powder and place in the first cup. Break one of the half tablets into several small pieces and place in the second cup. Place the unbroken half tablet in the third cup.

Add 40 mL of room temperature water to the first cup. Begin timing as soon as the water is added. Record the time it takes for the bubbles to stop forming. Repeat this process for each of the other cups.

Notice that you used the same kind of cups, the same amount of effervescent tablets in each cup, and the same amount of water in each cup. The same method of timing was used, and the water temperature was the same each time. The things that were kept the same are known as **controls**. The only thing that was different was the particle size. In this experiment, particle size is the **variable**.

	Reaction time
Fine powder	
Small pieces	
Whole piece	

Which half reacted at the fastest rate?

Which half reacted at the slowest rate?

Part II: In Part I you did an experiment to see how particle size affected the reaction time of effervescent tablets and water. The particle size was the variable. This time, plan your own experiment to determine the effect of temperature on reaction time, using three different water temperatures, ice water, room temperature water, and almost boiling hot water. Your teacher will boil some water and pour it into your cup. Write your procedure. Let someone else look at your procedure before you begin to see if you thought of everything you need to do. Remember, the temperature will be your variable. Think of the controls you want to be sure to keep the same in all three tests.

	Reaction time
Ice water	
Room temp. water	
Hot water	

Which half reacted at the fastest rate?

Which half reacted at the slowest rate?

There are things (called the controls) you kept the same in the three cups. What were the controls?

There is one thing that was different (called the variable) in the three cups. What was the variable?

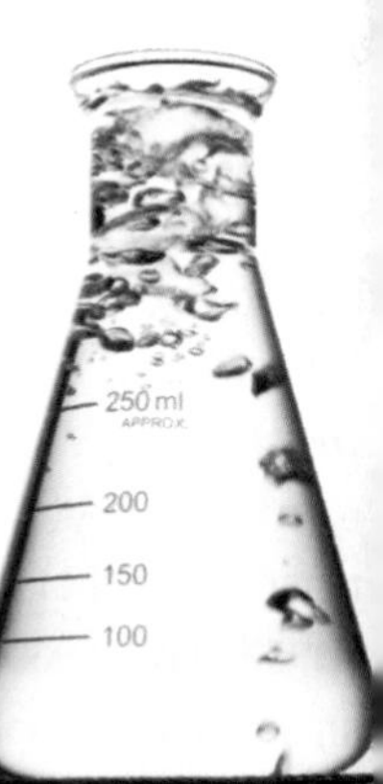

The Science Stuff

Substances sometimes undergo **chemical reactions** and change into different chemicals. Chemical reactions can take place quickly or slowly. Size of particles is one factor that affects the rate of a reaction. Smaller particles react faster than larger particles.

Chemical reactions occur as the molecules and other particles move around in the solution and bump into each other. The chemicals involved in a chemical reaction must be very close to each other to allow them to bump into each other. When the antacid chemical is in a finely ground state, the water molecules can get very close to more of the molecules of the antacid. When the antacid is in big pieces, the water molecules can only get close to the antacid at the edges.

In the first experiment, you kept everything in the three cups the same as much as possible except the size of the effervescent tablet particles. The things that are kept the same are known as the **controls**. These include the same amount and temperature of water added to each cup, same amount of effervescent tablets, same kind of cup, same timer, and same timing method.

Another factor that affects the rate of a reaction is the temperature of the chemicals. The molecules and other particles that are involved in the chemical reaction move faster if they are hot. They move slower if they are cold. The heat energy also causes the molecules to move farther apart, so they move throughout a wider area. The antacid molecules and the water molecules bump into each other (and react chemically) more often when they are moving faster and farther.

When you plan an experiment, you should include an **investigative problem** and a **procedure** (which tells how you did the experiment, and identifies the **controls** and **variable**). Generally a written report of an experiment will also include a **hypothesis, procedure, results,** and **conclusions**. A fellow student should read your written report about the experiment and try to analyze what you did right and what you did wrong. This is similar to what is known among professional scientists as a **peer review**.

Dig Deeper

Sometimes chemical reactions are sped up with the help of catalysts. Use a reference book to look up the term catalyst and give some examples of ways in which catalysts are used.

Time-release capsules allow certain medications to be released gradually and not all at once. What would be the advantage of slowly releasing a medication into your body? Find out more about time-release capsules.

Making Connections

Iron metals rust as they combine chemically with oxygen. Rust is the same thing as iron oxide. Iron is a shiny dark gray color and can be made into a magnet. Iron rust is a dull reddish color and cannot be made into a magnet. Steel wool is made up of fine strands of iron (or an iron alloy) and will rust much faster than a block of iron. The rate of the reaction is affected by the size of the pieces of iron. A block of iron will only form rust on the surface of the block. The steel wool has much more surface area in contact with oxygen and will turn into rust quickly.

One of the most important engineering processes developed in the United States is that of steel production. The chemical reactions that separate iron from iron ore occur in a hot blast furnace. They could occur at lower temperatures, but the reactions would take much longer and the quality of the metals would not be as good. The production of iron and steel products could not keep up with the high demand for them without carefully engineered methods for producing very high temperatures.

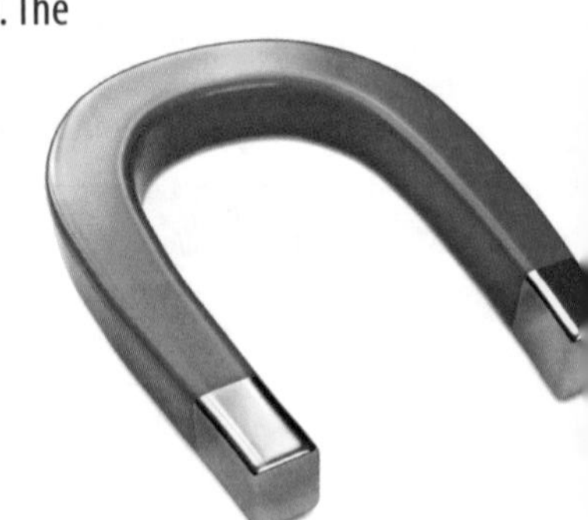

What Did You Learn?

1. Do all chemical reactions occur at the same rate?
2. Explain how the reactions you observed were made to occur at a faster rate.
3. In order for two molecules to react chemically with each other, do they need to bump into each other?
4. Why will small pieces of effervescent tablets react faster than one big piece?
5. A rapid chemical reaction occurs when zinc metal is placed in a container of sulfuric acid. Will the reaction occur faster if the zinc is divided into several small pieces?
6. Which would have more surface area — an apple that is cut into two halves or an apple that is cut into four quarters?
7. Explain why adding heat causes most chemical changes to react faster.
8. Give an example of an industry where very high temperatures are needed to cause a chemical reaction to proceed better.
9. What were the controls in Part II?
10. What was the variable in Part II?
11. What is the procedure in an experiment?
12. What is a peer review?
13. Can you do one experiment in which you test the effects of particle size and the effects of temperature at the same time?

The surface area of an object is how much exposed area it has.

Understanding Phase Changes

Think about This

Joe put a soft drink can in the freezer to cool his drink, but then he forgot about it. The next day, he discovered his soda had frozen, broken through the can, and made a sticky mess in the freezer. "Oh no," he complained. "I thought things would shrink when they got colder and expand when they got hotter."

Where did Joe go wrong in his logic?

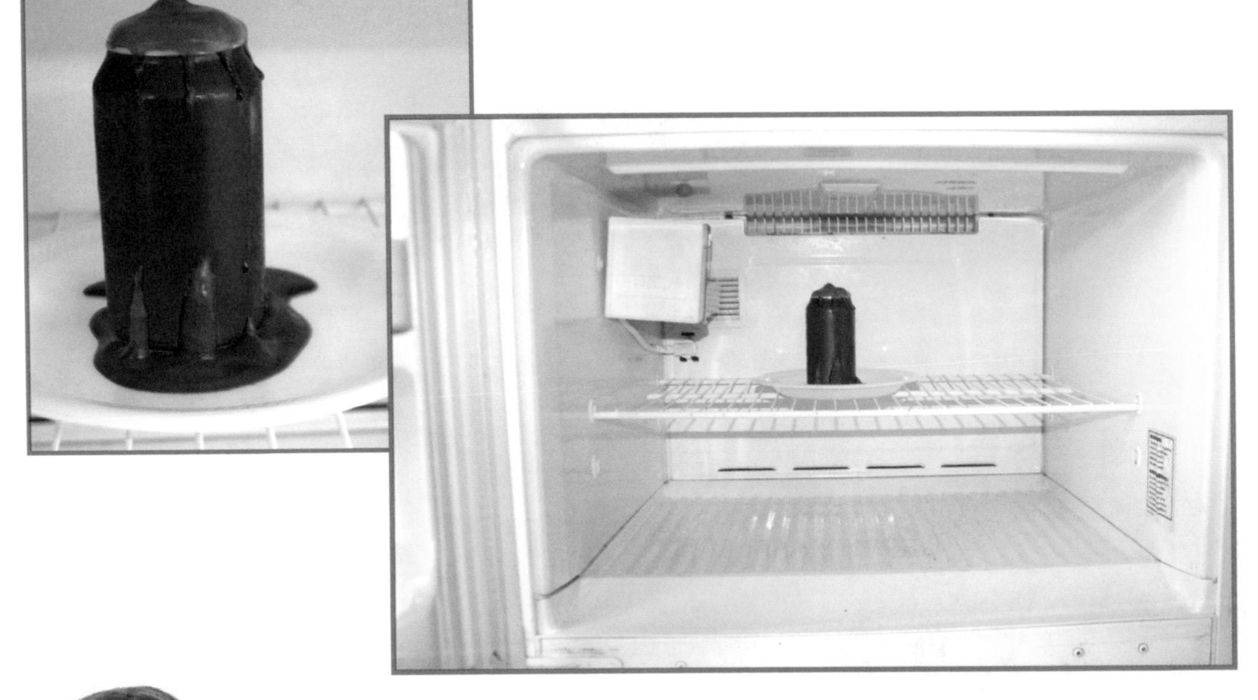

The Investigative Problems

What are the differences in a solid, liquid, and gaseous phase?

How close together are the molecules of a solid, liquid, and gas?

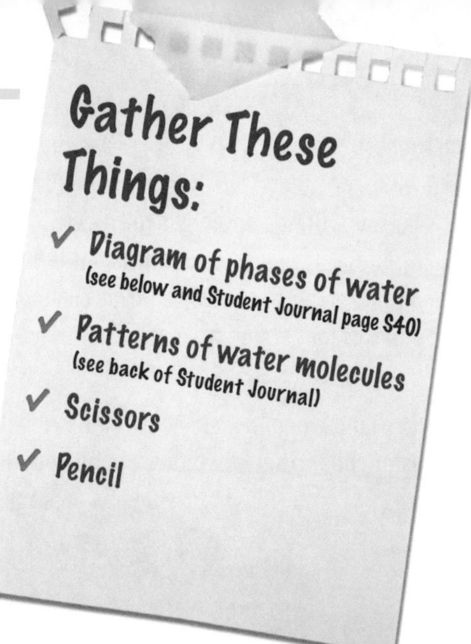

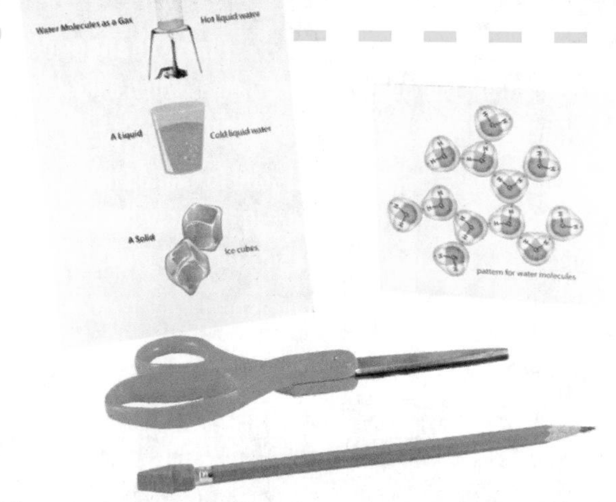

Three Stages of Water Molecules

Procedure & Observations

Using the diagram in the student book, place the cutout molecules together to represent the crystalline structures of molecules of water in its solid form.

Move up the page and arrange the same cutout molecules to represent water in the liquid form. Move up the page and arrange the same cutout molecules to represent water in its gaseous form.

Draw in the molecules beside each phase as they looked in the models. Label the changes that occur as you move up the drawing, using the terms **melting** and **evaporation**. Now start at the top of the drawing and move down, using the terms **condensation** and **freezing**.

Your drawings should contain some important information about the three phases of water.

As you move up the page (from solid to liquid to gas), energy is absorbed. As you move down the page (from gas to liquid to solid), energy is released. Write this information on your drawing.

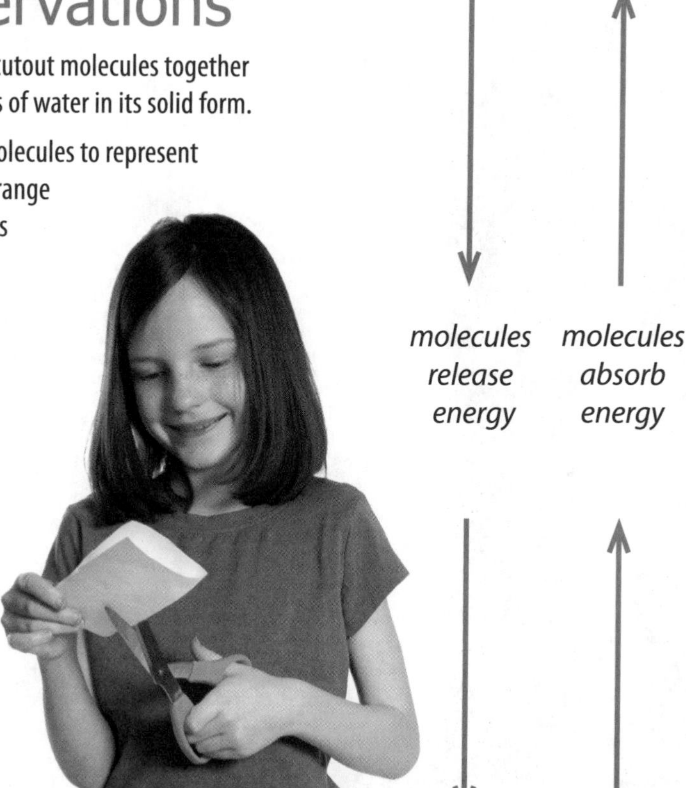

molecules release energy

molecules absorb energy

A Gas

Gaseous water in the form of steam

Hot liquid water

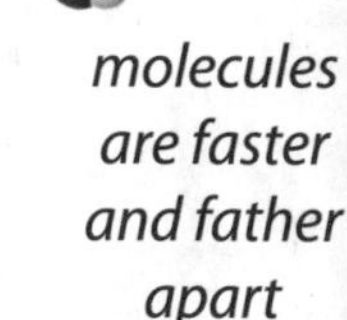

molecules are faster and father apart

A Liquid

Cold liquid water

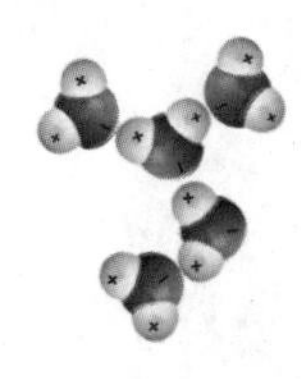

A Solid

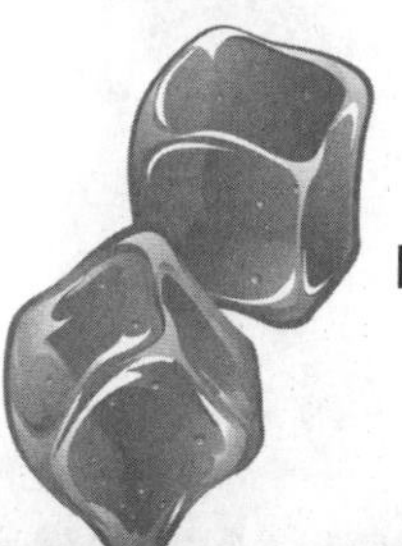

Ice cubes

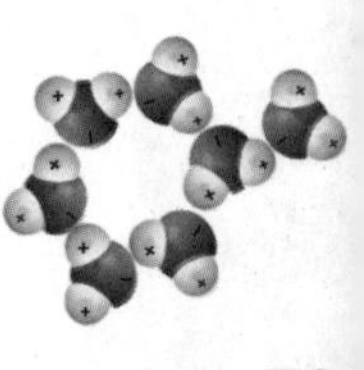

The Science Stuff

Phase changes occur when solids, liquids, or gases change from one phase into another, as when a liquid changes into a gas. All phase changes are physical changes, not chemical changes.

Some form of heat energy must be added for a solid to change into a liquid or for a liquid to change into a gas. Whether it is the temperature in the air or heat that is applied from a flame, there is an input of heat. As heat is added to a substance, it tends to expand and get bigger.

The process can also work in the opposite direction as heat is given off. When a gas is cooled, it gives off heat and its molecules get closer and closer until it turns into a liquid. As the liquid gives off more heat, the molecules move closer together until a solid state is reached. The motion of the molecules is slower in a solid. In fact, they may only be vibrating in place. Molecules never completely stop moving. So, as heat is removed from a substance, it gets colder and it tends to contract and get smaller.

Water is different from almost every other chemical in that it is bigger when it is frozen than when it is a very cold liquid. Because of the way water molecules attract each other, hexagonal shaped openings are part of the ice. Remember Joe's dilemma in "Think about This"? His logic was correct, but he failed to realize that the freezing of water is an important exception. Except for this, the other ideas about phase changes are true of water, too.

Making Connections

Railroad tracks are placed end to end, but spaces are left between them to keep the tracks from expanding and buckling in hot weather. Sidewalks are laid down with cracks between blocks of cement, so that the cement will have room to expand when it gets hot. Bridge engineers must design ways in which the metals and other materials in bridges can expand and contract as temperatures change. Designing railroads, sidewalks, and bridges does not involve phase changes, but this does illustrate how substances expand and contract according to whether they are absorbing heat or giving off heat.

Almost any solid, liquid, or gas will expand when heated, and shrink when it is cooled. Remember that molecules move faster and farther when they are heated; they move slower and get closer together when cooled.

Dig Deeper

You might not think molecules are moving around in a solid, but they are, even if the movement is just vibrating in place. Take a square of chocolate candy, place it on some aluminum foil, and put it in a pan on the stove. An adult will need to gradually heat it on a low setting until it just changes from a solid into a liquid, and move it away from the heat and give it a few minutes to cool. Place the chocolate and aluminum foil in a refrigerator for a few minutes if it doesn't return to a solid state. See if you can give a scientific explanation for why a square of chocolate candy will change from a solid to a liquid when you heat it, and why it will change from a liquid to a solid when you cool it.

Do some research on how bridges are engineered to allow the metals to expand when they get hot and contract when they get cold. Include diagrams and drawings to illustrate some of the various methods that are used.

What Did You Learn?

1. How close are molecules of a gas to each other compared to how close they are in a liquid?
2. How close do molecules of a liquid get to each other compared to how close they are in a solid?
3. Are phase changes physical changes or chemical changes?
4. What happens to the movement of molecules when heat energy is added to a substance?
5. What happens to the movement of molecules when heat energy is given off by a substance?
6. There is an important exception to the general rule that solids shrink when they are frozen. What substance gets bigger when it freezes?
7. Why are cement sidewalks made with cracks between the blocks?

The Race to Evaporate

Think about This

Your teacher will spray perfume on a warm surface at the front of the room. Raise your hand when you first smell the perfume and keep your hand in the air. The next person to smell the perfume should also raise his or her hand and keep it in the air. Continue for about five minutes or until everyone has detected the perfume. How far had the perfume molecules moved after five minutes? Do you see a pattern in how the molecules moved?

William Thomson (Lord Kelvin) was a mathematical physicist and engineer. He is known for developing the Kelvin scale of absolute temperature measurement.

The Investigative Problems

Which substance evaporates the fastest?

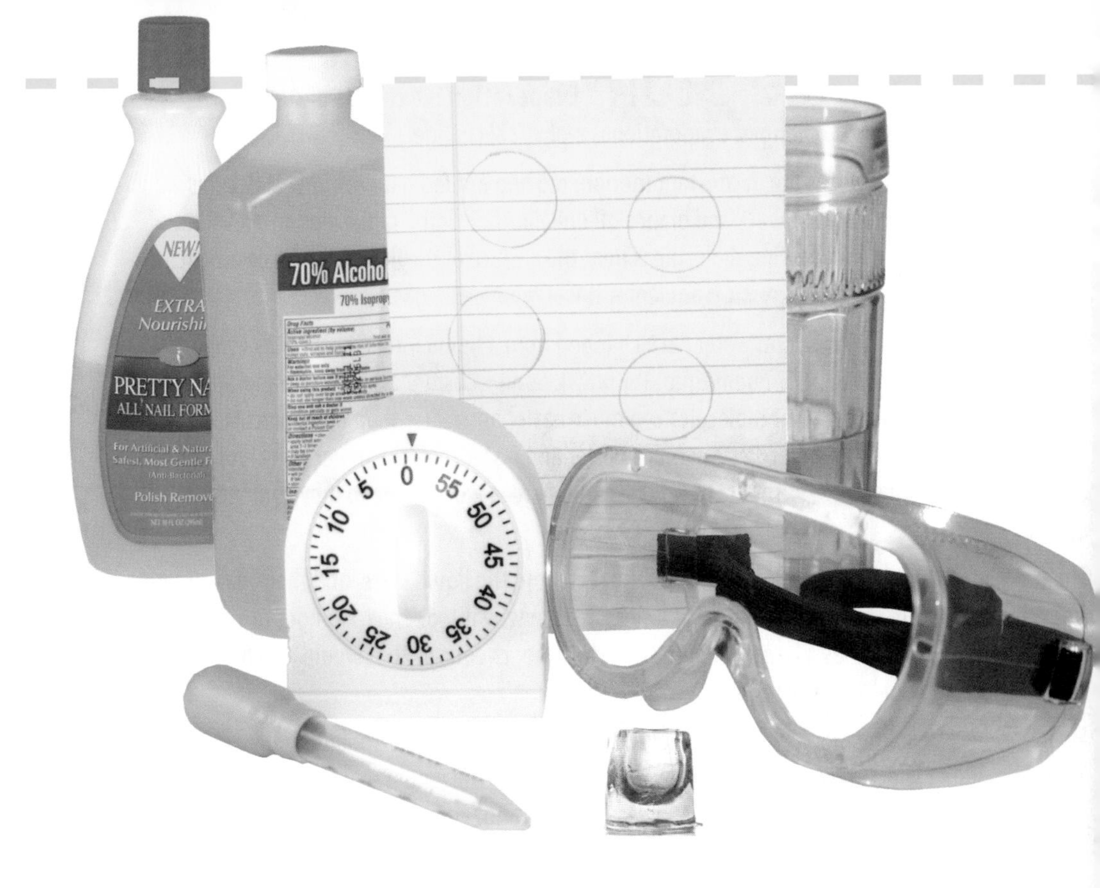

Procedure & Observations

Take the paper with the four circles and place it in front of you. Now place the piece of ice on one circle and label it "ice."

Wear safety glasses for the rest of this activity. Acetone and rubbing alcohol could be harmful to your eyes if they accidentally got into them. Your teacher will place two drops of acetone, two drops of alcohol, and two drops of water on the circles. Label the circle where each chemical is placed. You will be given one substance at a time so you can time each one more accurately.

Observe and note the time as soon as the substance is no longer visible. Record the time it takes for each substance to evaporate.

	Evaporation time
Ice	
Fingernail polish remover w/ acetone	
Rubbing alcohol	
Water	

Which one evaporated the fastest?

Rank the evaporation in order from the fastest to the slowest.

Have the teacher put two drops of rubbing alcohol on the back of your hand. Blow or fan over the rubbing alcohol until it evaporates. Does your skin feel cooler where the alcohol had been?

Put two drops of water on the back of your other hand. Blow or fan over the water until it evaporates. Does your skin feel cooler where the water had been?

Did you notice a difference in how cold your skin felt where rubbing alcohol had evaporated and where water had evaporated?

The Science Stuff

Evaporation is a change of state from liquid to gas. Like all phase changes, evaporation is a physical change.

In the liquid state, molecules are in constant motion, moving around haphazardly. Molecules may collide head-on or hit sideways and glance off one another. Some molecules are traveling faster than others. If a fast-moving molecule is near the surface of the liquid, it may be able to overcome the liquid's **surface tension** and enter the air above as a gas.

The acetone, rubbing alcohol, and water each changed from a liquid into a gas as they evaporated, but they evaporated at different times (or rates). Some substances tend to evaporate faster than others, because their molecules can move easier and faster than the molecules of other substances. Acetone molecules can move easier and faster than the molecules of rubbing alcohol or water. Substances without strong **cohesive forces** will be able to move around easily.

The piece of ice had to first change from the solid state into the liquid state, and then change from the liquid state into the gas state. That is one reason why it took the ice longer to evaporate. (However, since it was a larger amount than the other chemicals that would also be a reason why it took longer to evaporate.)

Evaporation is a cooling process. The faster evaporation occurs, the more cooling there will be. Therefore, under the same conditions, the evaporation of acetone would remove more heat from your skin than rubbing alcohol or water. However, alcohol and water also evaporate fast enough to produce noticeable cooling. The opposite of evaporation is **condensation**, where a gas changes into a liquid. Condensation is a warming process.

Both evaporation and condensation take place at the surface of a liquid. Boiling is similar to evaporation, but the bubbles of gas form below the surface during the process of boiling.

Making Connections

An effective way to speed up the drying of a wet shirt is to hang it in a place where there is sun and a breeze. Heat and movement of air increase the rate of evaporation.

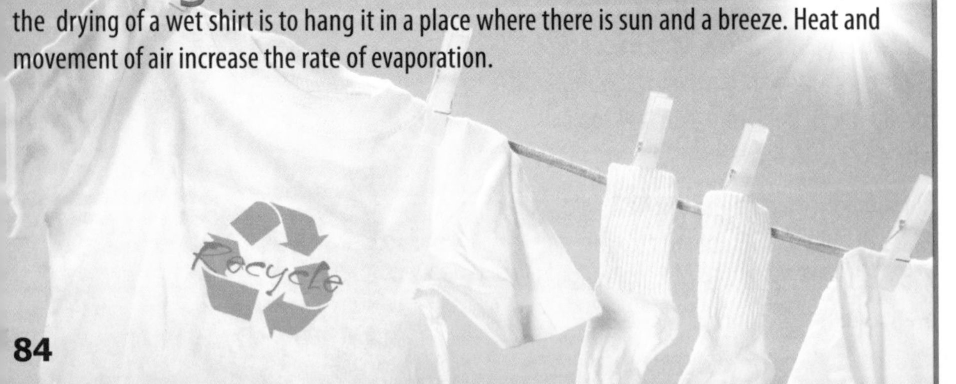

What Did You Learn?

1. Where did each substance go when it evaporated?
2. In what form was each substance after it had evaporated?
3. Are the substances you started with still in the room?
4. What are some of the factors that affect how quickly a substance evaporates?
5. Why does it take ice longer to evaporate than it takes an equal amount of water to evaporate?
6. In your investigation, the ice was not an equal amount compared to the other substances. Why would this indicate that you had added an additional variable?
7. Is evaporation a cooling or a warming process?
8. Explain how sweating helps to keep your body cool if you are running on a hot day.
9. Does evaporation occur at the surface of a liquid or below the liquid's surface?

Dig Deeper

Do some research on how sweating helps your body stay cool and may help prevent a heat stroke when you do strenuous work on a hot day. What is the first aid for a heat stroke?

Do some research on the science of cryogenics. Try to find how some substances behave when they are extremely cold.

Find some information about *absolute zero* and the scientist, Lord Kelvin, who introduced the term.

Narrative: Evolution Is Just a Theory

As you begin to do activities, it is easy to get terms like facts, theories, and explanations mixed up. All terms used by scientists should have specific meanings. Read the following narrative and see if you can tell why this is important.

"Hey, Dad, why did you just hit yourself in the head?" Joey asked.

"I was expressing a little frustration at those people being interviewed on TV," Dr. Houston laughed.

"Why? Did they do something bad?"

"No, but they're getting a lot of scientific terms all mixed up. I wish I could help them get a few things straightened out." Dr. Houston taught chemistry at the university and was a stickler for using terminology correctly.

"Like what?" Joey persisted.

"The School Board is trying to decide whether or not to approve the new high school biology curriculum. Those people on TV are expressing their opinions. The gentleman in the green shirt thinks evolution should only be taught as a theory. The lady in the red outfit is saying that evolution is a scientifically proven fact."

"So, who is right?"

"They are both having misunderstandings about what science is and some of the other terms they are using," Dr. Houston said.

"I thought everybody knew what science is."

"Most people think they do, but actually many people don't. For example, science is not about proving things. It is mostly about finding explanations for things that have been observed and seeing if the evidence agrees with the explanations," Dr. Houston explained, putting the emphasis on explanations and evidence.

"The other problem is that they don't understand the scientific meanings of theory and fact. To make things even more confusing, evolution has more than one definition," Dr. Houston continued. "One definition for evolution is simply that living things change over time. That is an easy thing to observe, so in that sense, evolution can be considered a fact. Evolution can also mean that all living things evolved from the same one-celled ancestor. This cannot be observed and cannot be considered a fact. The lady keeps switching back and forth between the two definitions, so it's frustrating to listen to her."

"Isn't it OK to say that evolution is just a theory?" Joey asked.

"The gentleman who is saying evolution is just a theory is trying to say evolution is just one possible explanation for where life came from. However, he isn't using the word theory in the same way a scientist would. When scientists use the term theory, they mean an explanation that is supported by a great deal of evidence from many different sources."

"So, what's wrong with saying evolution is a theory?"

"It gives the impression that there is strong evidence that all life came from a common ancestor. I don't like to call Darwinian evolution a theory, because there is no clear evidence that all living things, including man, evolved from the same original cell."

"Does evidence prove that something is a fact?"

"Not necessarily. Think about this. It's a fact that I'm wearing a gray suit and a red and black-stripped tie; and I'm 5'11" tall. Now suppose a policeman comes on TV and announces that someone fitting this exact description has just robbed a nearby store. The evidence provided by the store camera shows a masked man that looks like me. The video is a fact, but it doesn't prove that I was the robber, does it?"

"No, sir! You would never rob a store."

"Actually, the term scientifically proven fact is not something learned from doing an experiment. A fact is basically the same thing as making an observation or taking a measurement or collecting data. You don't need to do an experiment to prove that I'm wearing a red and black-striped tie. You just observe me and my tie."

"Another misconception is thinking that if you collect enough evidence, a theory can change into a fact. A theory doesn't become a fact except in those rare cases when the explanation becomes an observation. There can be many explanations for one set of facts. An explanation is strong or weak, depending on the kind and amount of evidence there is. By the way, a theory never changes into a scientific law."

"What's a scientific law?"

"A scientific law states a regular repeating pattern that has been observed over and over and is often expressed mathematically. Boyle's Law and Newton's First Law of Motion are examples of scientific laws.

"Look, Dad," Joey said, "The guy in the green shirt is pounding the table and yelling 'just a theory.' The lady in the red outfit is shouting 'fact, fact, fact.'"

"Joey, I hope you learn some lessons from this. First, people should never try to make their points by yelling. Second, both sides of a debate should be sure they understand the terms they are using. Third, allowing debates is a good thing. Many evolutionists claim Darwinian evolution is a fact because they don't want it to be debated or challenged in any way. I believe all available evidence and all logical interpretations of the evidence should be considered. Ignoring evidence is no way to do science."

Express an opinion: What do you think people mean when they talk about the "fact of evolution"?

How would you respond to someone who said evolution is a fact?

What are three things to remember about debating something with another person?

Chart of Densities

Substance	Density (g / cm³)
Ethyl alcohol	0.81
Ice	0.92
Water (at 4° C)	1.00
Aluminum	2.7
Zinc	7.1
Iron	7.9
Copper	8.9
Lead	11.3
Mercury	13.6
Some pine woods	0.5

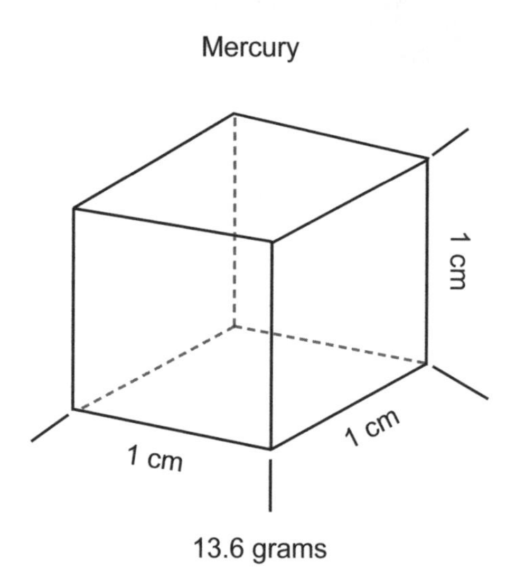

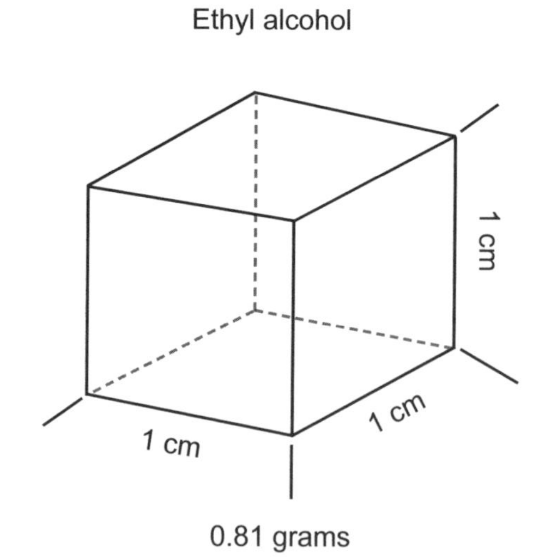

A cubic centimeter box full of mercury would weigh 13.6 grams while a cubic centimeter box full of ethyl alcohol would weigh 0.81 grams (not counting the weight of the box).

An ice cube will float on water because its density is less than one. A cube of aluminum will sink in water because its density is greater than one.

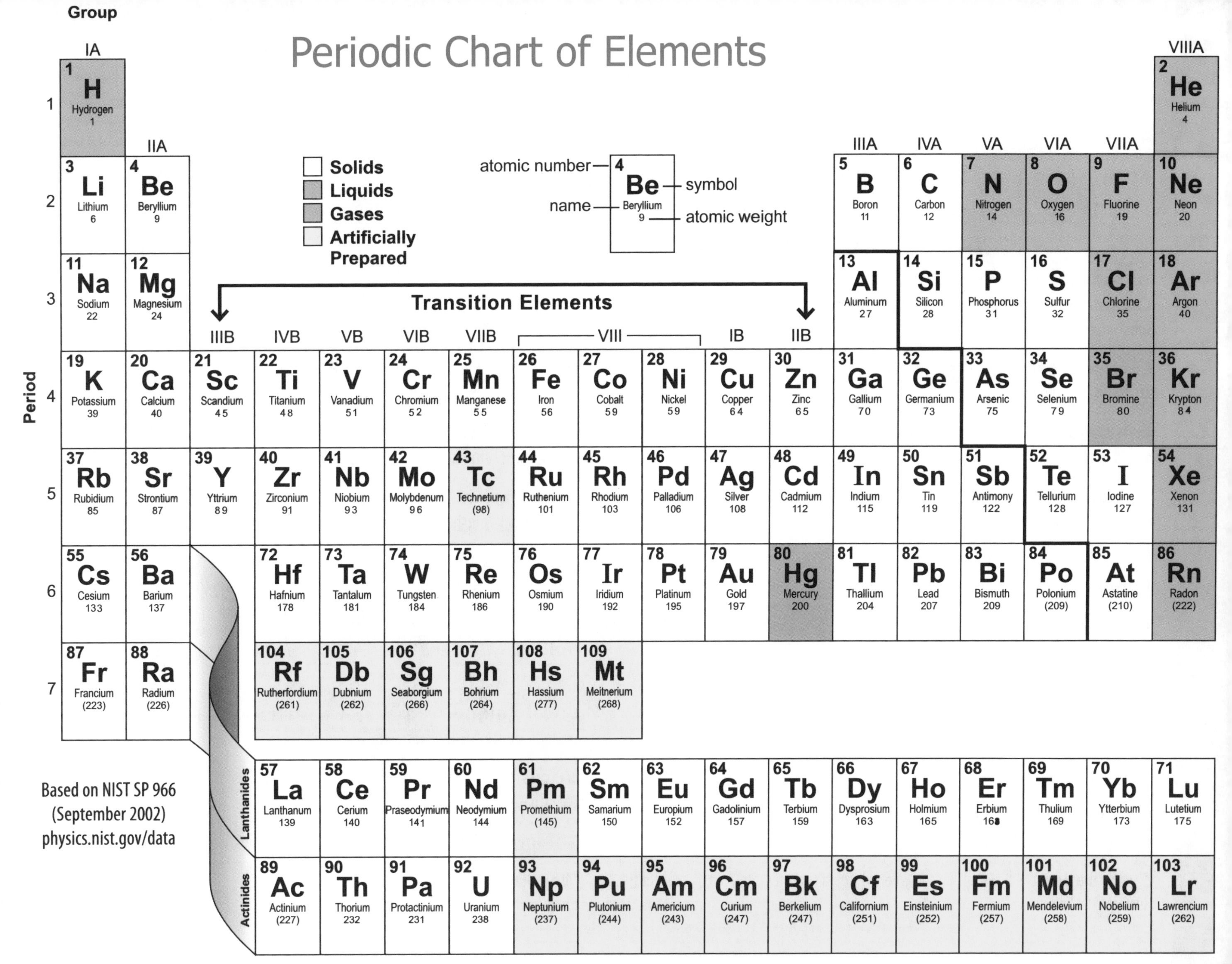
Group
Periodic Chart of Elements
IA
IIA
IIIA
IVA
VA
VIA
VIIA
VIIIA
Period
1
2
3
4
5
6
7
Solids
Liquids
Gases
Artificially Prepared
atomic number
symbol
name
atomic weight
4 Be Beryllium 9
Transition Elements
IIIB
IVB
VB
VIB
VIIB
VIII
IB
IIB
1 H Hydrogen 1
2 He Helium 4
3 Li Lithium 6
4 Be Beryllium 9
5 B Boron 11
6 C Carbon 12
7 N Nitrogen 14
8 O Oxygen 16
9 F Fluorine 19
10 Ne Neon 20
11 Na Sodium 22
12 Mg Magnesium 24
13 Al Aluminum 27
14 Si Silicon 28
15 P Phosphorus 31
16 S Sulfur 32
17 Cl Chlorine 35
18 Ar Argon 40
19 K Potassium 39
20 Ca Calcium 40
21 Sc Scandium 45
22 Ti Titanium 48
23 V Vanadium 51
24 Cr Chromium 52
25 Mn Manganese 55
26 Fe Iron 56
27 Co Cobalt 59
28 Ni Nickel 59
29 Cu Copper 64
30 Zn Zinc 65
31 Ga Gallium 70
32 Ge Germanium 73
33 As Arsenic 75
34 Se Selenium 79
35 Br Bromine 80
36 Kr Krypton 84
37 Rb Rubidium 85
38 Sr Strontium 87
39 Y Yttrium 89
40 Zr Zirconium 91
41 Nb Niobium 93
42 Mo Molybdenum 96
43 Tc Technetium (98)
44 Ru Ruthenium 101
45 Rh Rhodium 103
46 Pd Palladium 106
47 Ag Silver 108
48 Cd Cadmium 112
49 In Indium 115
50 Sn Tin 119
51 Sb Antimony 122
52 Te Tellurium 128
53 I Iodine 127
54 Xe Xenon 131
55 Cs Cesium 133
56 Ba Barium 137
72 Hf Hafnium 178
73 Ta Tantalum 181
74 W Tungsten 184
75 Re Rhenium 186
76 Os Osmium 190
77 Ir Iridium 192
78 Pt Platinum 195
79 Au Gold 197
80 Hg Mercury 200
81 Tl Thallium 204
82 Pb Lead 207
83 Bi Bismuth 209
84 Po Polonium (209)
85 At Astatine (210)
86 Rn Radon (222)
87 Fr Francium (223)
88 Ra Radium (226)
104 Rf Rutherfordium (261)
105 Db Dubnium (262)
106 Sg Seaborgium (266)
107 Bh Bohrium (264)
108 Hs Hassium (277)
109 Mt Meitnerium (268)
Lanthanides
57 La Lanthanum 139
58 Ce Cerium 140
59 Pr Praseodymium 141
60 Nd Neodymium 144
61 Pm Promethium (145)
62 Sm Samarium 150
63 Eu Europium 152
64 Gd Gadolinium 157
65 Tb Terbium 159
66 Dy Dysprosium 163
67 Ho Holmium 165
68 Er Erbium 168
69 Tm Thulium 169
70 Yb Ytterbium 173
71 Lu Lutetium 175
Actinides
89 Ac Actinium (227)
90 Th Thorium 232
91 Pa Protactinium 231
92 U Uranium 238
93 Np Neptunium (237)
94 Pu Plutonium (244)
95 Am Americium (243)
96 Cm Curium (247)
97 Bk Berkelium (247)
98 Cf Californium (251)
99 Es Einsteinium (252)
100 Fm Fermium (257)
101 Md Mendelevium (258)
102 No Nobelium (259)
103 Lr Lawrencium (262)
Based on NIST SP 966
(September 2002)
physics.nist.gov/data

GLOSSARY

Acids — A group of chemicals with a sour taste; release positively charged hydrogen atoms in water solutions.

Adhesion — A force that attracts different kinds of particles or materials to each other.

Atom — The smallest particle of an element that can still be considered that element.

Bases — A group of chemicals with a bitter taste and a slippery feel; releases a negatively charged particle composed of a hydrogen atom and an oxygen atom in water solutions.

Chemical change — The changing of one substance into another substance or substances.

Chemical formula — An abbreviation used to show the kind and number of atoms in a compound.

Cohesion — A force that attracts the same kind of particles or materials to each other.

Compound —Substances with different characteristic properties from the elements that make them up; can be broken down during chemical reactions into simpler substances.

Condensation — A phase change where a gas changes into a liquid.

Density — A physical property of a substance; found by dividing its mass (or weight) by its volume.

Distillation — A way of separating liquids in a mixture based on differences in boiling points.

Elasticity — Property of a solid to return to its original size and shape after being stretched or compressed.

Elements — Substances that do not break down by such chemical or physical means as heating, exposure to electric currents, treatment with acids, etc.

Evaporation — A phase change where a liquid changes into a gas; occurs at the surface of a liquid.

Hydrogen bonds — A weak bond that occurs between a slightly positive hydrogen and the slightly negative end of a polar molecule.

Indicator — A chemical that changes color in the presence of an acid or a base.

Mass — The amount of matter in an object; often used interchangeably with weight at sea level.

Mixture — The combination of different elements and compounds that are not chemically bonded together.

Molecule — Two or more atoms held together by chemical bonds.

Neutralization (acid-base) — When the positively charged hydrogen atom from an acid combines with the negatively charged oxygen-hydrogen particle from a base to form a water molecule; the other particles form a salt.

Physical change — A change in which a substance changes its physical properties without turning into a new substance.

Scientific law — Descriptive or mathematical statements that show regularly observed relationships or patterns in nature, but don't give an explanation for the observations.

Scientific theory — An explanation for something that has been observed in nature that is supported by a large amount of evidence.

Solubility — The ability of a substance to dissolve in another substance, usually a liquid.

Subatomic particles — The very small particles that make up an atom, such as protons, neutrons, and electrons.

Viscosity — A measure of the internal resistance of a fluid to flow; a fluid that flows easily has a low viscosity.

Volume —The amount of space an object occupies.

Weight —How hard the earth pulls on an object.